FLYING WITHOUT WINGS

FLYING WITHOUT WINGS

NASA LIFTING BODIES AND THE BIRTH OF THE SPACE SHUTTLE

MILTON O. THOMPSON AND CURTIS PEEBLES

SMITHSONIAN INSTITUTION PRESS
Washington and London

Copy editor: Lorraine Atherton
In-house editor: Ruth Spiegel
Designer: Janice Wheeler

Library of Congress Cataloging-in-Publication Data
Thompson, Milton O.
Flying without wings: NASA lifting bodies and the birth of the space shuttle / Milton O. Thompson and Curtis Peebles.
p. cm.
Includes bibliographical references and index.
ISBN 1-56098-832-0 (alk. paper)
1. Lifting bodies (Aeronautics)—Research—United States—History. 2. Space shuttles—Research—United States—History. I. Peebles, Curtis. II. Title.
TL713.7.T49 1999
629.47′1—dc21 98-48711
CIP

British Library Cataloguing-in-Publication Data is available

Manufactured in the United States of America
06 05 04 03 02 01 00 99 5 4 3 2 1

♾ The recycled paper used in this publication meets the minimum requirements of the American National Standard for Information Sciences—Permanence of Paper for Printed Library Materials ANSI Z39.48-1984.

CONTENTS

PREFACE

This book had its origins a decade ago and has seen both triumphs and tragedy. One day in the eighties, Milton O. Thompson came to Jack Kolf, a former X-15 flight planner, with a stack of papers. They were parts of a manuscript on Thompson's experiences in the X-15 program. Kolf typed and edited the text, adding paragraphs, fixing the grammar, and taking out the cuss words. Kolf's wife, June, and eldest daughter, Kathy, also worked on the text. The finished book, *At the Edge of Space,* was published in 1992 by the Smithsonian Institution Press and proved highly successful.

June Kolf, with four published books to her credit, started calling Milt "One-Book Thompson" and urged (nagged) him to do a second book on the lifting bodies. Thompson was in an ideal position to tell their story. Dale Reed was the idea man for these vehicles. He had seen the possibilities of the lifting bodies and had pushed for construction of a lightweight manned vehicle. It was Thompson, however, who made the program go. As a test pilot, he had lent his support to Reed's ideas, worked on the design of the M2-F1, then put his life on the line to fly a strange-looking, unproven vehicle. There were many at NASA who doubted the lifting-body concept. Thompson, Reed, and a small group of engineers set out to create a whole new type of aircraft.

Their efforts came at an important historical turning point in aerospace technology. Between the late forties and the early sixties, there was an assumption about how manned spaceflight would be accomplished. It was seen as a step-by-step process. Experimental aircraft would go progressively faster and higher until, at some future date, one would leave the atmosphere and go into orbit.

Indeed, in the fifties that was how events had developed: the X-1 and D-558-II had reached supersonic speeds, the X-2 raised the limits to Mach 3 and over 100,000 feet, and the X-15 was designed to achieve Mach 6 and reach the fringes of space. The final step would be the Dyna-Soar space glider. Launched by a rocket booster, the Dyna-Soar would go into orbit, then make a horizontal landing on the Edwards lake bed, just like the experimental aircraft before it.

But before the dream could become reality, events intervened. Changing space policy, internal Air Force squabbles, and the pressure of the U.S.-Soviet space race brought an end to Dyna-Soar. Manned spaceflight used capsules. These were very different vehicles from the experimental aircraft the old hands at Edwards had been used to. One *rode* a capsule rather than flying it. The astronaut was not a pilot but a lab rat. A capsule followed a ballistic trajectory, like a missile warhead. The capsule's path was determined by gravity and air resistance, and there was little the astronaut could do to change the landing point. Once the mission was over, the capsule would splash down, to float helplessly until the Navy arrived to save the "Spam in the Can."

The old hands wanted nothing to do with it. They were *pilots.* A lifting-entry spacecraft, like the Dyna-Soar, could be flown during the reentry. Like an airplane, it developed lift during reentry and used aerodynamic control surfaces to change its flight path. Its pilot could control the vehicle, down to a precise landing on a runway. The lifting-entry spacecraft could then be refurbished and launched again. Space travel would have the routine and flexibility of an airline.

In retrospect, the demands of space were greater than had been envisioned during the fifties. Rather than a continuation of the development of higher and faster experimental aircraft, the space program represented both a technological and psychological leap into the unknown. A long period of development was necessary before the great leap to a lifting-entry spacecraft could be made.

This was the importance of the lifting bodies: they showed it was possible to build a vehicle that could both survive the fires of reentry as well as remain stable in low-speed flight and make a controlled horizontal landing. The effort was long and difficult; nearly twenty years would pass between the first tentative flights of the M2-F1 and space shuttle *Columbia* thundering aloft.

With the urging of Kolf and others, Thompson started work on the lifting-body book. He had the storyteller's ability to convey to his audience the adventure, triumphs, setbacks, and fun of being a part of this effort. Thompson's writing technique was to make a list of subjects and events he wanted to cover. He would then write out the text by hand, printing it letter by letter, without organizing it into chapters. By the summer of 1993, Thompson had finished about 75 percent of the material he wanted to cover.

Milton O. Thompson died on Friday, August 6, 1993. That evening he was scheduled to be honored at a special dinner. He was to be awarded his second NASA Distinguished Service Medal and NASA's Exceptional Engineering Achievement Medal. Instead, the dinner was a wake. As it turned out, I was among the first to know of his death. That day I was at the Edwards Air Force Base History Office doing research. Dr. Jim Young came out of his office and told us that Thompson had died of an apparent heart attack. It came as a tragic surprise to all of us, as Thompson was only sixty-seven at the time.

Thompson's lifting-body book languished for the next few years. His family and friends wanted to see it in print, but the unfinished text was not in a form that could be sent to a publisher. Jack Kolf typed up Thompson's handwritten notes, then turned it over to Peter Thompson, one of Milt's sons. It remained with him until the family was contacted by Dr. Gregory "Hoss" Buchanan. Milt Thompson had been one of Buchanan's childhood heros, and he later said, "It was a supreme achievement to me to work toward this completion of Milt Thompson's second book." Eric and Peter Thompson agreed that he could look for someone to finish it. Buchanan spent a year writing letters, arranging meetings, and making phone calls, trying to get Thompson's last literary effort out of its holding pattern.

His efforts finally succeeded when Jack Kolf referred him to Dr. Richard P. Hallion, the Air Force historian. Hallion gave him my name and a glowing endorsement. Buchanan called me on February 1, 1997, and I agreed to work on the project.

Thompson's text arrived on February 5. It covered the background of the NACA and NASA, Thompson's activities in 1962 and 1963, the Paresev, the M2-F1, the development of the heavyweight lifting bodies, his flights in the M2-F2, fragmentary text on space shuttle development, and various notes on areas he wanted to cover. That material was organized into chapters 1 through 8 and chapter 10. Where material needed to be added, particularly in chapter 10, I used published interviews and reports from Thompson's technical archives, along with interviews with his family and coworkers. I wanted this to be Thompson's book.

There was a major gap, however. Thompson had written almost nothing on the lifting-body flights that took place after he retired from test flying, although his outline indicated several areas he wanted to cover. I used his outline, along with published accounts and documents, to write chapter 9. Because of the circumstances, I felt it would be best to do that chapter under my own name. The same is true for chapter 11, which covers events following Thompson's death.

Recent years have seen a return of interest in the lifting-body concept. There had long been debate among aerospace engineers about the relative advantages of winged versus lifting-body designs, regarding such issues as vehicle weight and internal volume. For a number of reasons, a winged design was selected in the early seventies for the space shuttle, rather than a lifting-body shape. Although the lifting bodies had shown that a space shuttle was feasible and they led to flight research in support of the shuttle program, their concept fell out of favor. The late eighties saw a resurgence of interest in lifting bodies. The initial proposal was for a small vehicle for space station crew rotation and resupply, which would allow cost savings and more flexibility over the shuttle. Interest grew, and other uses were proposed.

Today, the early work done by Thompson, Reed, and others on the wooden M2-F1 has led to two new lifting-body programs. The X-33 is a technological demonstrator for a lifting-body single-stage-to-orbit booster, which holds the potential to reduce greatly the cost of putting a payload into space. The X-38 is designed to test a space station crew return vehicle, which would be used should an on-board emergency require an evacuation. The shape is a direct copy of the X-23 and X-24A lifting bodies, first flown in the late sixties and early seventies. The production version could carry a crew of six and make an automated landing on Earth, using a steerable parachute. The X-38 could also lead to an operational vehicle to take crews to and from the space station. That would fulfill the role envisioned when the lifting-body concept was first proposed four decades ago.

In his lifetime Thompson received many awards, but he was always quick to say he could not have done it alone. As editor of Thompson's text, I too am deeply indebted to many people—first, to Eric and Peter Thompson, Dr. Gregory "Hoss" Buchanan, and Dr. Richard P. Hallion, for selecting me to work on this project. It is both an honor and a responsibility. Thanks also go to Ken Szalai, director of the Dryden Flight Research Center, for his interest; Bill Dana, chief engineer, and Jack Kolf, former X-15 flight planner and NASA X-24B project manager, for answering my numerous questions and taking the time to go over the draft manuscripts; Dr. J. D. Hunley and the staff at the Dryden Public Affairs Office, for providing access to Milt Thompson's technical archives; Dale Reed, Tom McMurtry, Lee Saegesser, Roy Bryant, Rick Norwood, Dr. Jim Young, Cheryl Gumm, Dr. Ray Puffer, and the staff of the Edwards AFB History Office; Dr. Rick Sturdevant; Peter Merlin; Tony Moore; Robert Tieman, manager of the Walt Disney Archives; and Sue Henderson.

Curtis Peebles

1
THE ROAD TO SPACE

When the space shuttle *Columbia* came out of the reentry blackout on April 14, 1981, my eyes filled with tears. NASA had accomplished a lifting reentry and a fly-back to a precise landing on Rogers Dry Lake at Edwards Air Force Base. When I arrived at Edwards in 1956, I didn't foresee what I was about to be part of. The golden age of flight test was nearing its end. We were almost finished with the X-1 program; the X-3, X-4, and X-5 programs had ended; and the X-2 was in trouble. I thought to myself, "The X programs are all over, I've missed it all." Then came the X-15, the Dyna-Soar program, the Paresev, and the lifting bodies. I hadn't missed it after all.

NACA FLIGHT RESEARCH, 1915–1945

The National Advisory Committee on Aeronautics (NACA), the forerunner and nucleus of NASA, was established by Congress in 1915. American research in aeronautics had fallen far behind that of the European powers. While France, Germany, England, Italy, and Russia had all established national aeronautical laboratories, no such organization existed in the United States. When World War I began the previous year, it was reported that France had 1,400 airplanes, Germany 1,000, Russia 800, England 400, and the United States only 23. In a 1915 survey of American universities, it was discovered that only two offered classes in aeronautics. Aeronautics was a subject of curiosity, rather than of significant engineering interest.

Congress specified that the NACA was to "direct and conduct research and experiment in aeronautics." The intent was to ensure that America would no longer be caught unprepared for another war, as we had been for World War I. When the United States entered the war, we had no real air force. No U.S.-designed airplane saw combat, and we had to become competitive using technology gained by our allies during the early years of the war. Congress wanted America to be preeminent and a leader in aeronautics.

The NACA initially established the Langley Aeronautical Laboratory in Hampton, Virginia. Intended to make aeronautics a science rather than an art, the lab consisted of a series of wind tunnels and other facilities to test various concepts. In one of its first endeavors the NACA set about creating a series of airfoil shapes and then tested them to determine their advantages and disadvantages. A designer could then pick an airfoil optimized for high-speed flight or one that improved low-speed handling, or change the airfoil shape along the wing in order to meet the design goals. Much of that initial airfoil section information still serves as a bible for current aircraft designers.

The NACA ultimately expanded to include the Lewis Aeronautical Laboratory, the Ames Aeronautical Laboratory, and the High-Speed Flight Station, which later became the Flight Research Center, and then the Dryden Flight Research Center. The research areas of these other centers were generally more specialized than Langley's. Lewis was primarily a propulsion-system and engine research lab. Ames tended to focus on high-speed aeronautics, flight control systems, and aircraft handling qualities. Dryden was established primarily to specialize in high-speed flight with aircraft such as the early X series. Each of the labs had a flight research organization, but they too tended to specialize in areas within the individual lab's specialty. Langley, however, conducted research in all of the aeronautical disciplines. It was also the only lab initially involved in helicopter research.

Early on, the NACA became involved in the quest for ever faster speeds, an effort that would ultimately lead to space. In the decade following World War I, most American aircraft used air-cooled radial engines. They were simple, lacking radiators and plumbing (something the Navy found important given the rigors of carrier operations), but had high drag. The cylinders had to be exposed to the air for cooling. The large frontal area created drag, which severely cut into the plane's performance.

In 1926, the U.S. Navy Bureau of Aeronautics asked the NACA to see if a circular cowling could be developed to reduce drag without creating too much of a cooling problem. The NACA used a Curtiss P-1A Hawk fighter from the

Army Air Service for tests. The plane was modified with a cowling around its engine cylinders. The results were phenomenal: with a small amount of added weight, the Hawk's top speed was increased from 118 to 137 miles per hour, a 16-percent increase. In February 1929, the race pilot Frank Hawks added an NACA cowling to a Lockheed Air Express, which he used to set a nonstop speed record from Los Angeles to New York of 18 hours and 13 minutes. The cowling had increased the plane's speed from 157 to 177 miles per hour. Lockheed Aircraft sent a telegram to the NACA following the successful record attempt: "Record impossible without new cowling. All credit due NACA for painstaking and accurate research." The new cowling produced a savings to the aviation industry of over $5 million, more than the total cost of NACA research from 1915 to 1928.

The twin breakthroughs of the airfoil shapes and the cowling made the NACA's reputation in the twenties and thirties. In addition, the NACA also developed superchargers (which improved the performance of piston engines at high altitudes), propellers, and design criteria for the aircraft manufacturers. It was obvious, for example, that some design criteria were needed to define good flying qualities for an aircraft.

The shortcomings of the criteria for aircraft flying qualities became apparent with American involvement in World War II. Paul Bikle was involved in testing many of the World War II aircraft while working at Wright Field, and he indicated that many of the newly designed aircraft were actually unstable. Some went into operational use before they were modified, and when Bikle asked several of the operational pilots what they thought of them, the pilots replied that the airplanes flew fine. Bikle realized then that the pilots didn't know the difference between a stable and an unstable aircraft. They were focused on getting the airplane up and down without crashing. It didn't bother them if they had to push on the stick as the airplane slowed down. They just aimed the nose of the airplane and did what was necessary to keep it pointed where it was aimed.

The NACA also tested most new World War II airplanes in wind tunnels, making various configuration changes to improve their performance, and then flew them to verify the performance increases. NACA pilots gained a vast range of experience during that type of testing. The NACA also examined each aircraft for any flying-quality problems that may have been identified by the military test pilots. It attempted to solve problems as they were encountered, or if the problem was generic in nature, it would attempt to define a configuration change that would eliminate the problem. The more dangerous the problem, the more urgent was the search for a solution. That was the job of the NACA test pilot. He flew where others feared to tread. He did it in a controlled manner, however, and in most cases did it safely.

RESEARCH PILOTS

The research pilot for the NACA or NASA is a rather special pilot, to say the least. As a combined engineer, scientist, and pilot, he becomes involved over the years in almost every discipline of flight. The need for such a specialized pilot became apparent as soon as the NACA began flying in 1919, using a pair of JN-4H Jenny trainers. It came up against the basic difference between flight testing and flight research. I tend to describe flight testing as a procedure to determine whether something new works or doesn't work as intended. Flight research goes one step farther and attempts to determine or verify *why* it works or doesn't work as intended.

During the early X-airplane program, at least two aircraft of each type were produced, primarily to have a backup in case the first aircraft was lost in an accident or other unforeseen event. The first aircraft of each type was delivered to the Air Force for flight testing. The second aircraft was delivered to the NACA for flight research. The Air Force and NACA flight teams worked together during the flight program, exchanging data or information and supporting each other as necessary. The Air Force team would typically expand the flight envelope of the new aircraft as rapidly as possible with a minimum of instrumentation in the aircraft. The NACA team would follow up with an aircraft that was heavily instrumented to measure the many forces, moments, and pressures acting on the aircraft in flight. The Air Force was, according to my definition, conducting flight tests on a new aircraft to determine if it performed as intended. The NACA, with its heavily instrumented aircraft, was attempting to determine why it worked or didn't work as predicted.

The NACA would also measure pressures on various parts of the airplane to calculate lift and drag. Those flight data were then compared with the wind tunnel data to determine whether the small model tests in the wind tunnels were accurately predicting the flight characteristics of new aircraft. The comparison was important to ensure that the wind tunnels were providing accurate data to the designers of new aircraft. If the wind tunnels were suspect, they were either modified to improve their predictability or were calibrated to define and document their inherent predictive errors.

Many of those NACA pilots remained active research pilots throughout their entire careers. Some were still flying high-performance fighter testbeds at age 60 or more. Military test pilots are not afforded that luxury. To gain promotions to higher rank and responsibility, they must move on to other nontesting and nonflying assignments to round out their experience. Bill Dana, one of our Dryden pilots, flew research programs on everything from sailplanes to the X-15 rocket aircraft in thirty-two years of active testing.

Amazingly, only a few have died during research flying. In forty-seven years of high-speed research flying, Dryden lost only three pilots in accidents. The first NACA pilot to die in the line of duty was Howard Lilly, killed in May 1948 when the jet engine in his D-558-I research plane exploded on takeoff from the Edwards lake bed. Joe Walker was killed in June 1966 during a photo session when his F-104N collided with the XB-70. Richard Gray died in November 1982 during a T-37 proficiency flight. Only Lilly's death occurred during a research flight—an impressive record considering the kind of aircraft that have been tested at Dryden.

NEW FRONTIERS, 1946–1957

World War II changed both the NACA and the nature of flight research. Before the war, the NACA was a small, obscure agency that even some congressmen didn't know about. The war caused a tremendous growth in the agency—in the number of people employed, its responsibilities, and the number of centers. Advances in aviation technology also opened new frontiers.

By the late thirties, airplanes such as the P-38 were flying at speeds in excess of 400 miles per hour. Pilots began reporting that in dives the plane would buffet violently and shudder, the dive would become steeper, as the airplane nosed over on its own, and the controls seemed frozen. As the airplane reached lower altitudes, it would mysteriously nose up and recover. Other times, the airplane would continue its dive until it was torn apart.

As the speed increased, air began to pile up in front of the wing, increasing drag and reducing lift. As the air traveled across the top of the wing, it accelerated until it was traveling faster than the speed of sound. This formed shock waves that moved back and forth, disrupting the airflow and causing severe structural stress. The increased drag meant that the horsepower of an engine had to be greatly increased to produce even a small increase in speed. As a plane approached the speed of sound, the horsepower requirements seemed to approach infinity. The British aerodynamicist W. F. Hilton observed that the speed of sound loomed "like a barrier against future progress."

The sound barrier, as it was soon known, gained a mythical quality. It was suggested that a pilot's voice would get stuck in his throat or that at supersonic speeds time would start to flow backward. In the summer of 1946, Great Britain abandoned all attempts at manned supersonic flight. In a press interview, Sir Ben Lockspeiser, the Ministry of Supply's director general of Scientific Air Research, said, "We have not the heart to ask pilots to fly the high-speed models, so we shall make them radio-controlled." Several months later, Geoffrey de Havilland died when his DH 108 Swallow disintegrated during a high-speed flight. His death fueled the popular belief that the sound barrier was impenetrable.

As World War II was ending, the Army Air Forces, Navy, and NACA all realized that high-speed flight, as well as new technological developments (such as jet and rocket engines and the new swept and delta wing shapes), would require development of special experimental aircraft, the X airplanes. The Army Air Forces developed the X-1 rocket plane, in which Capt. Chuck Yeager made the first supersonic flight. The Navy developed two planes, the D-558-I, a straight-wing, jet-powered aircraft for speeds up to Mach 1, and the swept-wing, rocket-powered D-558-II, in which NACA research pilot Scott Crossfield made the first Mach 2 flight. That first generation of X airplanes was followed by several advanced X-1s with top speeds over Mach 2 and the X-2 for speeds up to Mach 3. Several jet-powered aircraft were also built to test different wing configurations. The X-3 had short, straight wings; the X-4 had swept wings and no tail; and the X-5 had wings that could be moved in flight to vary the sweep angle.

Following the transition from the NACA in October 1958, NASA developed some unique research aircraft, such as the M2-F1, M2-F2/3, and HL-10 lifting bodies. It developed and tested the first fly-by-wire aircraft without any mechanical backup and the first digital fly-by-wire flight control system for an aircraft. NASA developed the first transonic supercritical wing aircraft and the Himat (highly maneuverable aircraft technology) remotely piloted research vehicle. NASA has also modified many other aircraft for research on natural laminar flow control, extreme angle of attack controlled flight, integrated propulsion and flight control systems, and flutter. It has even intentionally crashed an unmanned airliner under controlled conditions to evaluate new safer fuels, seats, and materials.

Early in the X-airplane program, the NACA and the Air Force worked in a complementary fashion. The Air Force managed the flight test program while the NACA served in an advisory and consultant role. Yeager described their relationship in his book *Yeager.* The Air Force in the persons of Yeager and Jack Ridley pressed aggressively ahead with envelope expansion while the NACA advocated a more cautious approach, basing each new flight on the results of the previous flight.

This combined team approach worked reasonably well with the X-1, since the flight envelope was successfully expanded with no significant problems. With the advent of the X-1A, one of the higher-performance versions of the original X-1, the relationship began to break down. The cause was the difference between the Air Force philosophy of flight test and the NACA philosophy of flight research. As Yeager pointed out in his book, the NACA was, in his opinion, much too conservative. He and Ridley ignored NACA warnings about potential stability and control problems with the X-1A. As the NACA and others predicted, the X-1A departed from controlled flight at Mach 2.4 during a December 1953 flight. The aircraft began to tumble so violently that Yeager's helmet cracked the plastic canopy. Yeager recovered from a subsonic inverted spin and landed successfully.

The Air Force approach had some merit; it provided a quick return of data from the aircraft. The advance of aircraft technology was accelerating from the late forties to the early sixties. The Air Force was developing numerous new aircraft. Some of the types in production and test were the F-100, F-101, F-102, F-104, F-105, F-106, B-52, B-58, and B-66. The Navy was also buying thousands of aircraft, including the F11F, F8U, F4H, A3D, and A3J. Several of them had top speeds close to those of the X aircraft. It was a race to keep the X airplanes ahead of the production aircraft.

The benefits of that approach did not offset the potential loss of the X aircraft, however, since they were one- or two-of-a-kind aircraft developed specifically to determine the merits or capabilities of the new concept. If they were lost, they could not be replaced, and the production aircraft would lose the benefit of their data. Yeager's narrow escape in the X-1A demonstrated the danger associated with an overly aggressive envelope expansion effort.

An even more vivid example is the X-2 maximum speed attempt on its last flight in November 1956. The Air Force Flight Test Center chose to make one more attempt to exceed Mach 3 before transferring the aircraft to the NACA. Paul Bikle confided to me that the Air Force made a mistake by attempting a maximum speed flight with a new pilot, Capt. Milburn Apt. The Air Force proceeded with the flight, even though they knew it was potentially catastrophic, because they were convinced Captain Apt would not be able to fly a perfect profile on his first flight. Apt flew the mission exactly as planned, reaching a maximum speed of Mach 3.2. As he turned back toward Edwards, the plane went out of control. Apt was thrown violently about the cockpit. He separated in the escape capsule but was knocked unconscious when its parachute opened. Apt died when the capsule hit the desert floor. The X-2 was to have been turned over to the NACA the following week. Mach 3 had been reached, but there were almost no data on high-speed flight or aerodynamic heating to show for it.

The X-1A's loss of control and the crash of the X-2 had been caused by inertial coupling. A new phenomenon, inertial coupling was primarily a result of the new aircraft configurations, which confined all the weight within a short distance of the aircraft centerline. Everything, including the fuel, was packed into a small, high fineness ratio fuselage, with short wings and tails to minimize frontal area and drag. It was great from the aerodynamic performance standpoint, but it had some other insidious effects. The mass of the aircraft was concentrated close to the roll centerline, while the damping from the wing was minimized for the same reason. When the aircraft made a turn, its angle of attack (the angle between the airflow and the wing) would change. Under certain specific conditions of speed, altitude, and roll rate, the plane's stability would decrease and it would suddenly go out of control.

During the X-2 flight, Apt found himself going faster than he imagined and was well beyond the lake bed. Apt had been warned not to use more than 5 degrees of angle of attack above a speed of Mach 2.4. When he tried to turn back to the lake bed, the X-2 developed an uncontrolled roll, then pitched up 20 degrees, and went into a supersonic spin. As the X-2 slowed to subsonic speed, it pitched down and entered an inverted spin. The plane was tumbling in three axes.

Joe Walker had also encountered inertial coupling during an October 1954 flight of the X-3 research aircraft. When it was first encountered during high roll rates, it was assumed to be a characteristic of the X-3 and not a generic problem. We at the NACA had done considerable analysis in an attempt to understand it. That same month, several F-100 fighters were lost when they broke up in flight. One of the pilots lost was George Welsh, North American Aviation's chief test pilot. It appeared to be an inertial coupling accident, and we offered to assist North American in its investigation.

Our pilots and engineers, including Scott Crossfield, began a methodical investigation of the phenomenon in an F-100 (serial number 778) to determine if inertial coupling was indeed responsible for the accident that killed Welsh and then to attempt to define a fix for the problem. Crossfield flew forty-five F-100 flights to define the inertial coupling limits. On one research flight the yaw was so severe that a vertebra in Crossfield's neck was cracked. After an extensive fast-paced investigation, we confirmed that the F-100A had a catastrophic inertial coupling characteristic. A solution was essential. The fix was an enlarged vertical tail to enhance directional stability and roll damping.

The first inertial coupling flight had some anxious moments for Crossfield that had nothing to do with the test. He had taken off on his first F-100 flight when a fire warning light came on. Crossfield shut down the engine and glided back to the Edwards lake bed, making a dead stick landing. Having made the first successful dead stick landing in an F-100, he decided to end the flight with a flourish—rolling right up the NACA ramp and stopping in front of the hangar. What he didn't know was that the plane had a hydraulic leak. He pushed the brakes three times to slow the plane, but it was still rolling slowly forward, like a diesel locomotive. When Crossfield pushed the brake a fourth time it went straight to the floorboard; the hydraulic fluid was exhausted. The F-100 rolled through the open hangar doors and punched its nose through the hangar wall. Yeager never let Crossfield live it down, saying, "The sonic wall was mine; the hangar wall was Crossfield's."

Inertial coupling was the kind of flight investigation that NACA pilots and engineers were involved with to an increasing degree after 1955. When a generic problem of this type was encountered, it was usually left to the NACA to solve it with its research engineers, wind tunnels, and research pilots. They devoted a lot of expertise to determining the extent and nature of the problem and then did extensive analysis to solve the problem. The NACA was often criticized for being slow in solving this type of problem, but it was seldom wrong in its solution.

The NACA's success with the F-100A inertial coupling problem led the Air Force and Walt Williams of the NACA to establish an informal program to have the agency evaluate each of the Air Force's new fighter aircraft. One of each new fighter (an F-101, an F-102A, an F-104, and an F-105B) was offered to the NACA to conduct a similar independent evaluation along the lines of the F-100 effort.

The NACA's first F-104 was acquired in late summer 1956. It was the seventh production F-104A, serial number 55-2961. With its short wings and T-tail, the F-104 epitomized the features that caused inertial coupling. A marginal dogfighter against the other century-series fighters, it was designed only to get to high altitude fast and pass by the enemy at a high enough speed to avoid a dogfight and a retaliatory missile. It was never intended to be a dogfighter; instead, it was a greyhound built to nip at the butt of the pit bulls as it passed by.

During the evaluation of the F-104A, both the Air Force and Lockheed lost all of their instrumented aircraft, as well as several test pilots, and finally asked to reacquire the NACA F-104 to complete their inertial coupling test program on schedule. The NACA was reluctant to return its F-104, since it had invested a substantial amount of money and manpower to instrument the aircraft extensively. The Air Force finally agreed to have the NACA conduct the investigation. Joe Walker was the project pilot on that program. The NACA and then NASA flew that first F-104A for nearly twenty years in various research investigations and as a proficiency and support aircraft. The aircraft was ultimately delivered to the Smithsonian's National Air and Space Museum, where it still resides.

After that first F-104 was delivered to the NACA, a second F-104 (number 56-734) was acquired, initially to determine if natural laminar flow could be achieved at supersonic speeds using an optimum airfoil surface. The optimum airfoil surface was achieved by applying a fiberglass coating on the right wing of the aircraft, which was subsequently shaped and smoothed to achieve an almost perfect airfoil surface. Thermocouples were installed in the fiberglass layer to measure the surface temperature and in this manner determine whether the airflow over the surface was laminar or turbulent. Turbulent flow imparted more heat into the fiberglass than smooth laminar flow. Surprisingly, the investigation worked as conceived—laminar flow was achieved on the fiberglass-coated wing over the forward 25 to 30 percent of the wing. The standard left wing was also instrumented and revealed a much lower amount of laminar flow.

Neil Armstrong made the first flights in this investigation. I inherited the program several months later. I joined the NACA in March 1956 as a research engineer and became a research pilot in 1958. I began an apprenticeship, starting out with minor programs. The first aircraft I flew at Edwards was a T-33 in a performance test looking at fuel usage. The second was an F-51 that we used as a utility plane. I then moved up to some minor support programs and flew chase flights in the F-100 for other research aircraft.

The flight profile for the laminar flow investigation was relatively simple. Take off and head out alongside the supersonic corridor to the east, climbing en route to the Colorado River to achieve 35,000 feet. Make a left turn at the Colorado River and enter the corridor heading back toward Edwards while initiating a maximum power acceleration to Mach 1.6. At Mach 1.6, I would begin a slight climb while continuing to accelerate to Mach 2. I would generally achieve Mach 2 at 60,000 feet or more and then maintain Mach 2 until the fuel reached "return to base" levels. That would usually occur over Mojave or just south of Tehachapi. I would then make a 180-degree turn and head back to Edwards.

The return to Edwards was made in a descending decelerating turn in an attempt to maintain a constant Reynolds number during the descent. These were short flights. I had only 4,900 pounds of fuel, and that didn't last long in full afterburner. I made more than 100 flights in that program and accumulated only 35 hours of flight time. The altitudes were so high that I had to wear a partial pressure suit, which fit extremely tight, like a girdle around my entire body. There were tubes running down the arms and legs that would inflate if the plane lost pressurization and tighten up the suit to prevent my blood from boiling. It was extremely uncomfortable and hot but was a real pleasure to remove after a flight. Thank God for the full pressure suits that became available a couple of years later.

The same F-104 was used for another research investigation, of boundary-layer noise on a smooth polished nose cone that was substituted for the production nose cone, an experiment that provided data to compare with wind tunnel data for calibration purposes. At the Air Force's request, the aircraft was returned to be modified into a target drone. I was extremely upset by that decision. It was a beautiful aircraft, carefully maintained and polished. It was really depressing to realize that it would be blasted apart by a missile and its surviving parts would end up on the bottom of the Gulf of Mexico. What a crime. What a waste.

The years following World War II had been a period of fantastic advancements in aeronautical technology. The twin revolutions of the jet engine and supersonic flight changed the face of aviation beyond recognition. Before the X-1, supersonic flight had been an unknown land. Yeager's flight opened the new frontiers. As we explored these frontiers, we found that it was not enough to refine the existing technology, as the NACA had in the twenties, but rather we had to develop a whole new technology. Everything was changed—structures, control systems, airfoils, and materials. New aerodynamic demands, such as inertial coupling, required development of the flying tail, larger fins, and electronic stability augmentation systems.

Those frontiers had been opened at great cost. Between 1947 and the end of 1956, a total of fifty-five pilots, flight engineers, and crew members were killed in aircraft accidents at Edwards. Streets were named for test pilots killed in crashes—Fitzgerald, Seller, Wolfe, Mortland, and Popson—and the base itself was named for Glen Edwards, who was killed in a YB-49 crash in June 1948.

The base had also changed during those years. Pancho Barnes's Happy Bottom Riding Club was long gone, as was the old tarpaper housing area called Kerosene Flats. Similarly, the NACA's original windswept dusty hangar and makeshift offices were replaced by a state-of-the-art research center. By the time I arrived at Edwards, there was the sense that an era was ending. The NACA was still flying the X-1B and X-1E rocket planes, but their usefulness was diminishing. With the extensive preparations needed for each rocket-plane flight, we could make only two or three flights per month. With the F-104s, we could make that many flights per day, to nearly the same speeds as the X-1s. The X-1B and X-1E were finally retired when cracks were discovered in their fuel tanks. It did not make sense to devote the resources needed to repair these tired old birds. Their research activities were transferred to the F-104s.

TOWARD SPACE

On October 4, 1957, the Soviet Union launched *Sputnik 1,* the first manmade Earth satellite. A year later, on October 1, 1958, the NACA went out of business, replaced by the National Aeronautics and Space Administration (NASA). Although Dryden's activities were centered on aeronautical research, we had long had our eyes on space. The early X aircraft were seen as round 1, raising speeds from subsonic to Mach 3. As early as 1951, studies were under way on round 2, an X aircraft with a speed of Mach 6 and a maximum altitude of 250,000 feet—reaching the edge of space. Development of the X-15 was formally approved in 1954.

Because of problems experienced with the relationship between the Air Force and the NACA, their dual approach to the early X aircraft was modified in the X-15 program. A joint test team approach was used: a team of Air Force, NASA, and Navy personnel flew the aircraft in a single envelope-expansion program. This arrangement has continued for almost all of the new X series and research aircraft programs generated by the Air Force since that time. It has been used in testing the X-24A and B, the X-29, and the X-31, as well as in the F-111 TACT (Transonic Aircraft Technology) program and the AFTI (Advanced Fighter Technology Integration) program involving the AFTI-16 and AFTI-111 aircraft.

Round 3 was to be a space glider, launched by a rocket booster to nearly orbital speeds, that would maneuver during reentry and land like a conventional aircraft on a runway. On October 10, 1957, the Air Force's Air Research and Development Command consolidated a series of studies into a single project called Dyna-Soar (a contraction of "Dynamic Soaring"), only six days after *Sputnik* was launched and a year before the X-15 was delivered. That was how we worked at the time—as one project neared flight testing, we would be looking ahead to the next project.

In 1959, Neil Armstrong, Bill Dana, and I were assigned as "pilot consultants" to the Dyna-Soar program. We worked with Boeing, the Dyna-Soar prime contractor, on all aspects of the vehicle and its booster, offering recommendations on design of the different systems. The three of us spent a tremendous amount of time flying the Dyna-Soar simulator, working on the flight control and guidance systems.

Although the X-15 and Dyna-Soar were our primary focus at Dryden, space activities became an increasing part of our flight research activities with the F-104s. NASA needed to test the Mercury capsule drogue chute deployment. The drogue chute was to be deployed from the Mercury capsule at supersonic speed to stabilize and decelerate the capsule prior to deployment of the main recovery chute. NASA had no convenient way to test and demonstrate the operation of this chute.

Dryden proposed using a modified F-104 to make the test. We requested the loan from the Air Force of an F-104A that had been modified with a weapons rack under the belly of the aircraft. The weapons rack could be extended below the aircraft, allowing large rockets or missiles to be fired without endangering the aircraft. It was the only F-104 to be modified in this manner. We would use the aircraft to carry a bomb-shaped simulated Mercury capsule to the planned speed and altitude for operational deployment. The dummy capsule would contain instrumentation to measure the deployment loads and a drogue chute with its associated deployment hardware. The dummy capsule would weigh as much as the Mercury capsule, roughly 1,800 pounds. We calculated that we could launch the dummy capsule at 70,000 feet at a speed of Mach 1.2 or greater. It was an interesting program. The dummy capsule was intended to impact the Edwards bombing range and then be reusable for successive launches. The impact velocity was quite high, but the dummy capsule and its instrumentation were supposed to remain undamaged.

It was a real struggle to get the desired launch conditions of Mach 1.2 at 70,000 feet, but after launch, everything worked according to plan. On occasion, finding the dummy capsule was tricky, even though it was tracked by radar and optics. The problem was locating the exact impact point and then digging down 10 or 15 feet to retrieve the capsule. Joe Walker flew this series of flights and really enjoyed them.

The next program on this plane was the Alsor program to launch a large balloon to 1 million feet altitude using a rocket mounted on the extended launch rack. The purpose was to measure the density of the atmosphere at a location on the X-15 flight track that would allow researchers to calculate the true altitude, airspeed, and dynamic pressure of the X-15 after a mission. Without the atmospheric density, those quantities could not be calculated accurately at the extreme altitudes routinely achieved by the X-15.

The proposed operation would begin with a near-vertical launch of the rocket carrying the balloon. At peak altitude, estimated to be in excess of 800,000 feet, the nose cone would separate from the rocket and the balloon would be deployed. It would be inflated by a drop of water inside the balloon that theoretically would vaporize at the near-vacuum pressure at these altitudes. The balloon was quite large, 10 feet in diameter, and would be tracked by radar to measure the rate of descent, and from that calculate atmospheric density. It was a good idea, but we could not locate the balloon on any of the trial launches. We abandoned that program after three launches to altitudes in excess of 600,000 feet.

The next program was similar to the Mercury drogue program. The Air Force asked us to launch a simulated B-58 escape capsule at supersonic speeds to test the capsule's drogue chute. I was project pilot on that program. We made a number of successful launches, which ultimately culminated in a capsule ejection from a B-58 at supersonic speeds. For this test, a live brown bear was strapped in the capsule to verify that the ejection loads were survivable. The bear survived, but he really became nasty when they tried to put him back in the capsule for another test. He wasn't as dumb as a pilot!

Two other planes we used at this time for space-related research were an F-100 and a two-place F-104B, acquired from the NASA Ames Research Center. NASA headquarters had decided to concentrate all of its high-performance aircraft research programs at Dryden. Previously, each NASA center had conducted a wide range of research, on aircraft from helicopters to supersonic jets. The F-100 had been extensively modified for use as an in-flight simulator. Its flying characteristics could be varied to match almost any fighter aircraft, both real or imagined. We flew it for several years to simulate various X-15 characteristics as well as some U.S. supersonic transport flying qualities.

We utilized the F-104B to conduct some zero-g experiments, initially involving a small fuel tank designed to supply fuel continuously under zero-g conditions. It was not a simple problem to solve, since at zero g the fuel is floating around in the tank in various-sized blobs. Some interesting concepts were tested with this aircraft.

A second program using the F-104B involved an indirect viewing system for use on high-speed aircraft where windows may not be easily provided. At hypersonic speeds, the air temperature impinging on a window or windshield can exceed 2,000 degrees F, and special, extremely heavy glass is required to withstand such temperatures. If some indirect viewing system—using a periscope, for example—could be developed to enable the pilot to fly the aircraft, substantial weight and cost savings could be realized. The indirect viewing system that Dryden developed relied on a pair of periscopes originally developed for a military tank. They were installed in the rear cockpit and extended above the canopy behind the front cockpit. The view from the periscopes was outstanding, with a field of view in excess of 180 degrees laterally and approximately 60 degrees vertically. The only significant problem resulted from spacing the optic lenses farther apart than human eyes, which exaggerated the stereoscopic effect and caused some errors in depth perception when approaching touchdown. Other than that, the system worked well. We made steep simulated X-15 unpowered landings without any major problem. The system was uncomfortable to use, however, since the pilot had to keep his head and eyes pressed against the eyepieces. This became a problem during elevated g maneuvers.

The primary use by NASA of F-104s during the X-15 program was to simulate X-15 landings for pilot proficiency purposes. The X-15 pilots practiced approaches and landings at each of the emergency lake beds before each flight. This proficiency requirement dictated a minimum of two F-104 aircraft to provide the flight time for our three X-15 pilots. A third aircraft filled in during unanticipated down times for the primary proficiency aircraft. This X-15 pilot proficiency requirement became the driver for the subsequent acquisition of three new F-104N aircraft to be used for proficiency purposes only. They were originally numbered 011, 012, and 013 and were painted with bright orange and yellow markings for high visibility; however, the paint scheme was artistically applied.

PROFICIENCY AIRCRAFT FOR THE ASTRONAUTS

The F-104Ns were not the only colorful proficiency aircraft I was involved with. Walt Williams and I were on an aircraft traveling to Langley from Washington, D.C., in 1960 when the subject of proficiency aircraft came up. Williams was the Space Task Group deputy for operations at the time, responsible for astronaut selection, training, simulation, and numerous other aspects of astronaut care and feeding, and he wanted my opinion on what type of aircraft to provide for astronauts. They had been flying F-102 aircraft provided by the Air Force, but they wanted something a little more challenging. They were advocating F-104 aircraft. I immediately challenged that choice.

The F-104 was a dangerous and unforgiving aircraft. If you lost the engine, you lost the aircraft, and quite often you lost the pilot also. It was not a realistic proficiency aircraft. If NASA had acquired those aircraft for that purpose, it might have lost half of the original seven astronauts. Lockheed had lost a number of veteran test pilots in the F-104, and every other organization that flew it lost a wily veteran or two during transition. Besides, the astronaut proficiency aircraft was not intended to be solely a proficiency aircraft. It was to serve as an airborne taxi, support aircraft, and trainer. The F-104 was not an option in my opinion.

I recommended the T-38, a new hot-rod supersonic trainer that could double as a support and cross-country transportation vehicle. Williams and I argued for a short time but then moved on to another subject. I later learned that Williams and the Space Task Group had selected T-38s and issued a contract to Northrop for an initial batch. I was delighted to hear that, but I didn't think anymore about it for a long time.

One day, I was alone in the pilots' office at Dryden. A call came in from someone in the sales department at Northrop. He wanted to know what color to paint the astronaut T-38s that NASA had procured.

I will always consider that call to be one of the most momentous of my career. I had in my grasp the power to present the astronauts with some pink or purple aircraft. All I had to do was tell him the color and hang up. He would have followed instructions. He was not a manager or decision maker. He was simply asking for NASA's decision. I've despised myself for more than thirty years for not specifying a color and then hanging up. I could not have been caught or convicted. No one else was in the office, and anyone could have answered the phone. I was gutless. I could have been the most famous man in NASA. I could have been more famous for specifying a color than for flying to the Moon. I didn't do either, and that's probably why—no guts.

DIFFERENT ROADS TO SPACE

At Dryden we saw spaceflight as an extension of atmospheric flight. X aircraft would fly higher and faster, until finally they would go into space. The assumption went deeper than that, however. We believed we would *fly* into space. The pilot would have control of the vehicle during launch, and when the mission was over, he would land on the Edwards lake bed, just as we did in the X aircraft.

The Mercury capsule took a different road to space. Mercury came out of missile technology. The booster was controlled by its own guidance system; the pilot had no means of control during launch. The Mercury capsule would automatically carry out its mission. The astronaut was to sit quietly, like a good biological specimen. The landing would be by parachute. The capsule would splash down and the astronaut would be rescued by half the U.S. Navy.

Most of the old hands at Edwards didn't want anything to do with the space program, especially at first. They flew, they didn't ride. Yeager liked to say that an ape would make the first flight. There was a halfhearted attempt to have Dyna-Soar beat Mercury's schedule, but it was soon clear that this would not be possible. We did think that if Mercury ran into trouble, we might be the first Americans in space. By 1961, the Soviets had beaten us in the race to put a man in space, Kennedy had started the Apollo program to beat the Soviets to the Moon, and Dyna-Soar was in trouble and headed for cancellation.

The people at Dryden were an independent bunch, and we still believed in the value of a spacecraft that could fly to a controlled landing. By late 1961 we had started a number of projects aimed at filling the gap left by Dyna-Soar. They were done on a shoestring budget, and half the time we hid them away from NASA headquarters in case they disapproved.

Sometimes the participants in a major flight research project don't really grasp the importance of what they are doing or realize what it will lead to. Only in retrospect do they realize what they were a part of. The Paresev paraglider and the M2-F1 lifting body were small projects, but in retrospect we were opening the road to that first shuttle landing two decades later.

2
A TEST PILOT'S DIARY, 1962–1963

The years 1962 and 1963 were busy ones for me. During those two years, I was supporting the Dyna-Soar program as pilot astronaut, flying the first Paresev paraglider research vehicle, serving as consultant to the Manned Spaceflight Center in Houston on the Gemini parawing program (an outgrowth of the Paresev project), flying the first lifting body, serving as a chase and support pilot for the X-15 program, and finally, flying the X-15 as a project pilot. Much of this activity was overlapping and conflicting at various times. A sample of my activities during these years illustrates the variety of the flight tests and extensive travel involved.

This kind of flight research doesn't exist anymore. Today, a NASA research pilot would work on only one or, at most, two projects at a time. Nor would there be the opportunity to fly so many different types of aircraft. An Air Force test pilot can be checked out in only two or, at most, three aircraft at a time.

This period was also a dynamic time for another reason. The paraglider and lifting bodies were new concepts. They represented exciting possibilities for land recovery of spacecraft. At the same time, we had little in the way of practical experience with them. This created the need for actual free-flight testing of the concepts. By taking a simple approach, it was possible to undertake significant flight research with what amounted to homebuilt aircraft.

1962: BIRTH OF THE PARESEV AND DYNA-SOAR DUTIES

On January 25, 1962, I began the initial flight tests of the Paresev, which we had constructed to evaluate the potential capability of a parawing recovery system. The Manned Spaceflight Center was interested in the concept as a recovery system for the Gemini spacecraft. The parawing would potentially enable the Gemini to be recovered on land. In early March of that year, I attended a meeting at North American Aviation in Downey, California, to fly a Gemini parawing simulator. North American had won a contract to develop and demonstrate a parawing recovery system for the Gemini spacecraft.

Two days later, on March 7, Joe Walker (my boss at the pilot office) and I were in Houston to brief the Manned Spaceflight Center management on our Paresev program and our Lunar Landing Research Vehicle (LLRV). The LLRV would be developed by the Bell Aircraft Company to demonstrate the planned lunar landing in a piloted free-flight vehicle.

I had continued to fly the Paresev during ground-towed flights through January and February, and then on March 12 I made the first two high-altitude free flights to demonstrate the maneuverability and landing capability of this vehicle. The following day, I made two more free flights from high altitude. Then on March 14, 1962, I began the checkout of Bruce Peterson in the Paresev. Peterson had just been selected as a research pilot a week before, and this was his first research flight. He crashed on his very first ground tow.

On March 20, I traveled to Minneapolis, Minnesota, to fly the Minneapolis-Honeywell F-101 aircraft, which had a Dyna-Soar control system and side-stick controller to simulate its flight system. Some minor aircraft problems prevented me from flying the aircraft on the scheduled mission. I returned to Edwards the next day, following a stop in St. Louis to participate in a Gemini capsule design review.

I drove to Seattle on March 25 to spend a month supporting the Dyna-Soar development program at Boeing. Each of the three NASA pilot-consultants—Neil Armstrong, Bill Dana, and I—spent every third month at Seattle working in the simulator on reentry control techniques. While at Boeing, I was instructed to return to Minneapolis to fly the modified F-101 on April 3. This time I managed to complete an evaluation flight. From Minneapolis, I continued on to Worcester, Massachusetts, to be measured for a Dyna-Soar pressure suit. I returned to Seattle the following day to complete the month of duty.

These Seattle trips were not without their anxious moments. On one return trip I was booming down the freeway between Salem and Eugene, Oregon, in my 1957 Jaguar XK-140 roadster. I had finished my month of duty at Boeing, and I was eager to get back to Edwards. It was winter, and the road was covered with slush. As long as I didn't have to maneuver or change speed rapidly, I could make pretty good time driving in the slush. I was doing about seventy, kicking up a rooster tail of slush and water. The freeway in that location had two lanes in both directions with occasional paved crossover lanes in case a driver wanted to reverse directions. I was cruising in the outer lane behind another car traveling at about the same speed. All of a sudden, a puff of white smoke came out from under the car I was following, and the car began decelerating rapidly. The rate of deceleration was greater than what would be expected from brakes alone. Something had apparently seized up in his drive train. I turned to the left to avoid him, and I immediately began swerving around while still traveling straight down the highway. It was a classic slow-motion slide with the car rotating completely around. There were a number of cars on the freeway behind me that got a close-up look at my Ice Capade performance in my Jaguar. They whizzed past me wide-eyed as my car slid backward into the crossover lane and came to a stop in the center of the median.

I immediately checked my engine instruments, and everything looked good. I hit the accelerator and the engine responded normally, so I put it in gear and pulled back onto the freeway. Within seconds, I was passing the cars that had witnessed my spin. I had lost only a couple hundred yards of forward progress. There was a general look of disbelief on the faces of the cars' occupants as I passed them and waved. I was luckier than hell, but I didn't want to let them conclude that. I managed to smile and nod my head as though I had just accomplished a challenging display of driving skill.

As I climbed the mountain south of Medford, Oregon, it began snowing heavily. The highway patrol was stopping traffic several miles up the mountain to require motorists to install chains. I had no chains, so I waited until a big tanker truck came by and then pulled in behind him without lights. I made it through the road block and then on up and over the mountains into California. It was quite a trip in the middle of the night, but as I said earlier, I was in a hurry to get home.

While I was gone on these trips, my wife, Therese, had to make do with four young children. The kids knew I was a test pilot, but they also thought everyone's dad was a test pilot. There were five test pilots in my neighborhood in Lancaster, and Fred Haise lived down the street. They would come over to the house and discuss test flying. The kids were all quite young at this time, and it was not until later that they really understood.

I spent the last part of April and the first week of May 1962 in X-15 support activities. On April 30 I flew a chase mission on an X-15 flight. Then, on May 3, Bill Dana and I flew our C-47 up range to check on some of our X-15 lake beds to determine if they were usable. There were many more such flights in the years to come.

On May 9, I flew up to Monterey, California, in the C-47 to pick up Forrest "Pete" Petersen, the Navy X-15 pilot. Pete had gone up to Monterey to give a talk on the X-15 to the students at the U.S. Naval Postgraduate School. He had initially planned to fly up to Monterey in our F-51 support aircraft, but he managed inadvertently to nose over while taxiing out, damaging the propeller. As a result of that minor accident, we disposed of that F-51. We had not been flying it often and had planned to get rid of it anyway. The local U.S. Army Test Organization salvaged the aircraft and used it for a number of years as a chase aircraft for some of its higher-speed helicopters.

On May 18, I made the first air-towed flights in the rebuilt Paresev. Three days later came the Nellis fiasco. Neil Armstrong had been making practice X-15 landings on the lake bed in an F-104 when he damaged the plane's landing gear. He made an emergency landing at Nellis Air Force Base, closing the runway. I was picked to fly up in a two-seat F-104B and return him to Edwards. I had never flown the F-104B before, but Joe Walker assured me it flew just like the single-seat plane.

As soon as I lifted off and began wobbling across the sky, I knew Walker had not been completely forthcoming. Each of the early-model F-104s we flew had its own particularities. This one's handling was a lot looser than I had been used to. The fun continued when I arrived at Nellis. The tower didn't tell me about a strong crosswind. On my third landing try, I decided to use the landing flaps, which made the plane's handling poor. I forced the plane onto the runway, made a hard landing, blew a tire, and shut down the runway again.

The base operations officer met me as I walked in. He told me this was the second time that day the runway had been shut down. He calmed down a little after the runway was reopened. He then informed us that there were no F-104 parts at Nellis. Armstrong and I were stuck. Walker was mad as hell when we called to tell him about our problem. When he calmed down, he told us the C-47 was grounded, so no parts could be sent to fix my plane. He would have to send *another* plane out to get us.

While we were waiting, an Air Force officer inquired about what we planned to do with the two damaged F-104s. It turned out he was the squadron commander of one of the last F-104 units and was overjoyed by the supply of spare parts that had just fallen out of the sky. We tried to make it clear to him we were not about to let him loot our planes, but he was slow to get the message.

Bill Dana flew out in a T-33, and Armstrong, the operations officer, and I went out to watch. Dana landed so fast we thought he was not going to stop before running off the runway. The operations officer said something like "Oh, no, not again," Armstrong covered his eyes, and I just watched, transfixed. Dana managed to stop before going off the end of the runway, but the long-suffering operations officer had had enough. He begged us, "Please don't send another NASA airplane," and personally offered to find me a ride home.

Armstrong and Dana flew back to Edwards in the T-33, while I went back that night on an Air Force C-47 that was passing through. The ops officer apparently gave him gas just to get me away from Nellis and his much-abused runways. Armstrong, Dana, and I were outcasts at Nellis, and I didn't go back for a long time.

On May 23, 24, and 25, I made some weather and other support flights for the X-15 program. An attempt to fly the Paresev on May 24 was canceled. The last week of May, I was in Albuquerque, New Mexico, at the Lovelace Clinic undergoing an extensive annual physical examination.

On June 18, Bill Dana and I made a flight in a T-37. Neither of us had read the flight manual thoroughly. After landing we got an overheat light on one of the engines. We decided to shut it down. The problem was that neither of us could remember how to shut the engine down. It wasn't a normal throttle shutoff. We finally had to get the crew chief to shut it down, after we got into the chocks. That was my checkout flight.

The next day, I made my first flight in an L-19 that we had borrowed from the Army to tow the Paresev. I was back at North American in Downey on June 21 and 22 to attend a Gemini parawing meeting. On June 28, Neil Armstrong was towing me in the Paresev for the first time. He had never done any towing before. Primarily as a result of poor radios and miscommunications, Armstrong turned too tightly as we approached the edge of the lake bed. He rolled out of the turn headed south. I was still heading north in the Paresev, and the towrope was hanging slack between us. I had to release the towline and land in the sagebrush.

I flew weather chase and range-checkout flights in support of the X-15 program almost daily during the first half of July. July 19 started with a weather flight in a T-33 and ended with the LAX fiasco. De E. Beeler, deputy director of the NASA Flight Research Center, needed a flight to the TWA terminal at Los Angeles International Airport to catch an airliner to Washington, D.C. I was picked to fly him out in the T-33. We landed at LAX, but I had barely touched the brakes when both main wheels blew. It took more than an hour and a half to fly our C-47 out with two tires and fix the T-33, as it sat forlornly on the side of the runway. Among those delayed by the closed runway was Joe Walker, who was flying back from Washington, D.C. As his airliner touched down, he looked out and saw that one of *his* planes was the cause of the delay.

A few days after the LAX fiasco, Forrest Petersen left the X-15 program to assume command of a fighter squadron at NAS Miramar on July 23. He had to leave to continue his naval career. He later commanded an aircraft carrier, the USS *Enterprise,* and retired as an admiral.

On August 15 and 16 I attended a Gemini capsule design review at McDonnell Aircraft Company in St. Louis. I spent the remainder of the month flying the Paresev to obtain performance data. This included as many as four air tows per day.

From August 27 through September 7, I was at the Naval Air Research and Development Center at Johnsville, Pennsylvania, participating in a Dyna-Soar simulation in the center's centrifuge to determine how well a pilot could fly the booster into orbit. The results indicated that the pilots were quite capable of flying the booster, much to the surprise of the manufacturer. I was formally named as a Dyna-Soar pilot-astronaut on September 19 at an Air Force Association meeting in Las Vegas, Nevada. Neil Armstrong had been selected as a NASA astronaut two days before. Prior to his departure from Edwards, I checked him out in the Paresev, since he might eventually fly a Gemini capsule with a paraglider recovery system.

I spent the first two weeks of October in Seattle at Boeing, participating in the Dyna-Soar development program. During the latter part of October, I checked Gus Grissom out in our Paresev and also flew chase flights on the X-15. I was in Worcester, Massachusetts, on November 5 for another Dyna-Soar pressure-suit fitting. On November 7, I began a checkout of Bob Champine in our Paresev. The checkout was interrupted on November 9 because of an X-15 accident at Mud Lake. The X-15 landing gear had collapsed during slideout, severely injuring the pilot, Jack McKay, and severely damaging the aircraft. I had to fly the accident investigation team up to Mud Lake in our C-47. I completed Bob Champine's checkout the following week. I made a quick trip up to Seattle on November 20 and 21 for some Dyna-Soar publicity photos. I made another quick trip to Seattle on December 10 and returned three days later. On December 15, I made a dinner speech at an aviation club meeting in Palm Springs, California.

I was flying a weather flight on December 20 in one of our F-104s prior to an X-15 flight when the flaps malfunctioned. The aircraft began rolling violently, and I ejected. The F-104, which we had used for the Mercury drogue chute tests, dug a hole in the Edwards bombing range. Word of the accident spread quickly. My son Peter, who was then seven years old and in the second grade, recalled later that a teacher or the school principal came into the classroom and told him, "You have to go home. Your dad was just killed in a plane crash." As he walked home, he wondered when Dad would be home. He did not yet realize what it meant to have a father who was a test pilot.

On December 31, I was flying our C-47, searching for the F-104's ejection seat. Our personal equipment specialist wanted to recover the special emergency kit on the seat, since he had put a lot of man-hours into designing and fabricating it for our aircraft.

1963: THE FIRST M2-F1 FLIGHTS AND THE END OF DYNA-SOAR

I started 1963 with a trip to the Lovelace Clinic in Albuquerque on January 3, to be examined for any possible injuries that might have resulted from the ejection. The night I arrived, I was driving to my motel when a teenager pulled out of a side street directly in front of me. I hit him broadside and totaled both cars. I ended up in the hospital at Lovelace Clinic with a broken hand and a severe laceration on my upper lip. The next day, they examined me for any injuries that might have resulted from my ejection. They found none. Dr. Randy Lovelace, who examined me, somewhat seriously suggested that I tell my boss that my current injuries were a result of the ejection and that they had somehow been overlooked during my postejection physical. Lovelace was upset that I was returning to Edwards in worse shape than when I left.

On January 9, 1963, I spent the day at North American Aviation discussing the Gemini parawing program. On that same day, Walter Whiteside was accepting delivery of a new Pontiac Catalina convertible in Los Angeles. The car was to be used as a ground-tow vehicle for the M2-F1 lifting body that we were constructing. I was scheduled to be the project pilot of the M2-F1 to demonstrate the feasibility of the lifting-body concept.

January 15 and 16 were spent at the Ames Research Center in Mountain View, California, discussing the planned lifting-body flight program and reviewing wind tunnel tests of the M2 models. I spent the last couple of weeks of January rechecking Peterson out in the Paresev. He was scheduled to become the project pilot on the Paresev to enable me to concentrate on the development of the M2-F1 lifting body. In the meantime, the new Pontiac was delivered to Mickey Thompson's shop in Los Angeles for some performance improvements and structural modifications, which were completed during the last week of January. We now had a real muscle car.

I spent the last few days of January and the first week of February in Seattle at Boeing working on the Dyna-Soar. The remainder of the month was devoted to the M2-F1 simulator and routine proficiency flying at Edwards. The M2-F1 lifting-body construction was completed in the latter part of February, and on March 1, I made the first attempt to fly it while being towed by the Pontiac. I managed to get the M2-F1 airborne, but only for a few seconds at a time. The vehicle was uncontrollable. We immediately decided to test the vehicle in the full-scale wind tunnel (40 feet by 80 feet) at the Ames Research Center. I spent the next two weeks at Ames participating in the tests by sitting in the M2-F1 while it was in the wind tunnel and positioning the controls as force and moment information was being recorded.

The following week, I was undergoing the annual Dyna-Soar pilot's physical at Brooks Air Force Base in San Antonio, Texas. I returned to Edwards and spent the last week of March flying proficiency and support flights in our F-104 aircraft.

Our first successful ground-towed flight of the M2-F1 occurred on April 5, 1963. The control system had been reconfigured, and the vehicle flew surprisingly well. I made a number of flights behind the Pontiac on that day. We were ready to start the M2-F1 flight program, but requirements of the Dyna-Soar program interrupted our plans. All the Dyna-Soar pilots were requested to be at Boeing the next week for pressure-suit tests and publicity pictures. I flew the pilots up to Seattle in our C-47. The third week of April was devoted to proficiency and support flights at Edwards. During that week, I checked out in our new Aero Commander, used to support the X-15 ground stations, and the next day flew it to Albuquerque to deliver our chief of flight operations to the Lovelace Clinic for an examination.

I resumed flying the M2-F1 on April 19 and continued to fly it on ground-tow flights for the next three weeks. At the same time, I was flying weather and chase flights in support of the X-15 program. On May 13, I checked into the Lovelace Clinic for an operation to remove a polyp in my colon. Randy Lovelace performed the surgery. He was proud of his surgical technique, in which he cut through the stomach muscles parallel to the fibers in each layer to minimize the muscle damage. It worked. I was doing situps in bed two days after the operation.

I really enjoyed my recuperation in the hospital. I met a woman on the hospital sun porch who had undergone the same operation. She was a good friend of the doctor. Every afternoon during visiting hours, her husband would arrive with a cooler jug full of margaritas. The three of us would sit out on the sun porch and drink margaritas all afternoon. The hospital staff didn't really approve of this, but Lovelace was the boss. During my recuperation, I also volunteered to participate in some aero-medical tests that some of the research physicians were conducting. Lovelace was not too happy with the research doctors, since he felt that I should be resting. He contradicted himself, though, by inviting me to his home one evening for dinner. It was an excellent dinner, much appreciated after eating the bland hospital food. The next day I asked him if I could leave the hospital for a few hours each day to relieve the boredom. Initially he said no, but I reminded him that he had allowed me to leave the hospital for dinner at his home. He finally agreed that I could leave the hospital for three hours each day and informed the hospital staff of his decision.

Everyone assumed that I would leave during the day. I was a little smarter than that. I waited until 9 P.M. and showed up at the ward nurse's desk dressed and raring to go. At first she wasn't going to let me leave, but I showed her the doctor's order. He had failed to designate any specific hours. She finally let me go. I walked out of the front door and hobbled down to the nearest bar. My stomach still hadn't fully healed, but I could stand the pain. It was a lively place, and I spent the next three hours living it up. This developed into a nightly routine for the remainder of my ten days in Albuquerque. Lovelace didn't find out about my nights out until I checked out on May 23.

I took a few days of leave after returning to Edwards and then drove to Seattle to spend the first week of June at Boeing. I was grounded until a reexamination scheduled for June 10. I was cleared to fly again after the exam and then began flying proficiency flights in our F-104s.

In preparation for air tows of the M2-F1, we decided to get some operational experience for all involved parties by towing a glider with our C-47. We planned to use the C-47 as the tow plane for the M2-F1. On June 24, Don Mallick, Jack McKay, and I flew up to Tehachapi to pick up a sailplane. Mallick and McKay dropped me off at Tehachapi, then Fred Harris towed me in the sailplane back to Edwards with his Super Cub. We spent the next day installing a tow hook on the C-47, and the following day we began towing the sailplane. Everything went smoothly. When we completed our towing exercise, Mallick and McKay towed me back to Tehachapi with the C-47 and we returned the sailplane. Later that day, I also made a proficiency flight in our F-104.

I flew proficiency and support flights in our F-104s for the next couple of weeks. I flew the M2-F1 during ground tows on July 15 and 16. The next day, I flew a team of X-15 personnel up to Beatty, Nevada, in the Aero Commander to man the radar site in support of an X-15 flight. I continued flying proficiency flights in our F-104s through the remainder of July and on into the first week of August.

During this time I was informed by Paul Bikle, director of the Dryden Center, that I would replace Joe Walker as an X-15 pilot. I began X-15 ground school classes, simulator training, and landing simulations using our F-104 aircraft. My F-104 flying now became focused on X-15 approach and landing practice at each of the X-15 launch and emergency landing sites. I was also becoming more involved in other X-15 support activities, such as controlling X-15 flights from our remote tracking stations. I made three more trips in the Aero Commander to Beatty on August 9, 13, and 22 for this purpose.

In the middle of this rather hectic period, on August 16, I made the first free flight of the M2-F1 after being towed to altitude by the C-47. This culminated a long series of ground-towed flights. I had spent many, many early mornings roaring up and down the lake bed behind our souped-up Pontiac preparing for this first flight. Thank God, it was successful. Our second flight did not take place until August 28, but then in rapid succession we made five more flights in the next six days. The M2-F1 free flights were rather short, usually less than 4 minutes, but it required about half an hour to reach the desired release altitude while being towed by the C-47. It was a little like skiing without a ski lift: half an hour going up and 3 minutes coming down.

On September 4, I was in Worcester, Massachusetts, at the David Clark Company, being fitted for a new pressure suit for use in the X-15. I was becoming a steady customer. (Our Dyna-Soar pressure suits were also produced by David Clark.) I returned to Edwards via Grand Rapids, Michigan, to evaluate a new X-15 cockpit instrument panel being designed by Lear Seigler. For the next four weeks, I was immersed in X-15 training. I practiced X-15 landings at least every other day using our F-104s, but now I was focusing on the lake beds that I would use on my first X-15 flight. On my first flight I would be launched over the Hidden Hills dry lake, about 150 statute miles northeast of Edwards, and fly back to Edwards passing over two emergency landing sites at Three Sisters and Cuddeback dry lakes. Three Sisters was about 70 miles northeast and Cuddeback was 35 miles north-northeast of Edwards. Bill Dana, a fellow NASA X-15 pilot, used to describe simulated X-15 landings as "hurling oneself at the ground." I would hurl myself at the ground six to ten times on each flight.

I made several more M2-F1 free flights on October 7 and 9 and then returned to my X-15 training regimen and flying X-15 chase flights. I flew two more M2-F1 free flights on October 23, an X-15 chase flight on October 24, two more M2-F1 flights and an X-15 chase flight on October 25. On the following Tuesday, October 29, I made my first X-15 flight after flying a practice flight early that morning in an F-104. That was a *busy* week.

I flew more F-104 proficiency and X-15 chase flights the following week and then began preparing to check out Don Mallick, Bruce Peterson, and Chuck Yeager in the M2-F1. I made three air-towed free flights to verify that the M2-F1 was ready for the checkout flights. Then we began a series of ground-towed flights for the three pilots on November 12, 1963. These M2-F1 ground-towed checkout flights were interspersed by an attempt to make my second X-15 flight on November 19, which was subsequently aborted due to weather, and a successful flight on November 27. On December 2 and 3, Peterson and Yeager both made their first air-towed flights in the M2-F1. That concluded our M2-F1 flying for the year.

I spent the next two weeks in Dayton, Ohio, at Wright Field participating in some Dyna-Soar pressure-suit tests. While I was gone, Chuck Yeager was forced to eject from an NF-104A during a record altitude attempt. The plane was an F-104 modified with a rocket engine in the tail to boost it above 100,000 feet. Three were modified to train students at the Air Force Test Pilot School in the use of reaction controls at altitudes too great for conventional control surfaces to be effective. Yeager was badly burned on his face and neck from a fire in his pressure-suit helmet.

Yeager's crash occurred on December 10, 1963. It was not a good day for another reason. The Dyna-Soar cancellation was announced by Defense Secretary Robert S. McNamara that same day. All of us working on the project had known for months that it was in political and money trouble, but the problems had started early on.

McNamara originally wanted to take over the NASA Gemini program. In November 1962, he had suggested that Gemini become a joint NASA–Air

Force program with control being transferred to the Defense Department. When that idea did not fly with either NASA or the Air Force, his attention shifted to a separate space-station program to fly military experiments. By October 1963, McNamara was telling the Air Force that he was not interested in further Dyna-Soar funding. The Dyna-Soar was replaced by the Manned Orbiting Laboratory (MOL), a Gemini spacecraft attached to a lab module and launched by a Titan 3C. The MOL program limped along until 1969, when it too, like Dyna-Soar, was canceled without ever making a single manned flight.

I thought the cancellation of the Dyna-Soar was a pity. It was a solid research program, but it fell victim to political infighting within the Air Force. The aeronautical part of the Air Force was developing Dyna-Soar and had ideas for such missions as reconnaissance and using it as an orbital bomber. The Titan 3C booster was being developed by the space part of the Air Force, so there was a struggle for control of the program. I always thought that conflict was the reason for the end of Dyna-Soar.

I don't really know what would have happened to the overall U.S. space program had Dyna-Soar flown. We could have had routine operations in space years before the shuttle flew, but we were not far enough along in the early sixties to visualize the Dyna-Soar orbiting or repairing satellites. None of the proposed Dyna-Soar military missions were really taken seriously. The experience from the Dyna-Soar might have influenced the design of the space shuttle, so we could now have a much different kind of vehicle.

There were several new technological hardware developments during the Dyna-Soar program. A new inertial measuring unit coupled with a flight-qualified digital computer for guidance and navigation was developed, which was later used in the X-15 number 3. Another was a new auxiliary power unit that utilized hydrogen and oxygen as an energy source. A new Dyna-Soar pressure suit was developed that improved on the original X-15 suit. There was also new instrumentation for the pilot's display panel, skid landing gear that could be used on concrete or dirt strips, and a new side arm controller that was thoroughly optimized for both aerodynamic and exoatmospheric flight.

A major advance from the pilot's point of view was manual control of the boost phase using the booster flight control system. This was a significant philosophical change, since all boosters had previously been automatically controlled. There were many opponents of manual pilot control. Unfortunately, we pilots were never given the opportunity to demonstrate that capability.

An off-the-pad abort system was developed to save the Dyna-Soar spacecraft if a major problem developed on the launch pad before launch. This system included an abort rocket that was capable of boosting the Dyna-Soar to an altitude high enough for the pilot to make an unpowered approach to a skid strip constructed in the vicinity of the launch pad. This abort maneuver was extensively demonstrated by Neil Armstrong using one of our F5D aircraft configured to match the low lift-to-drag ratio of the Dyna-Soar vehicle.

The launch pad abort rocket was also planned to be used for an abort during the boost phase and also for deorbit. I wasn't too enthused about using it during the boost. I could visualize a situation with indications that a structural failure of some kind had occurred, or was about to occur. Usually when you encounter a problem in a new plane, you want to slow down, not increase your speed instantaneously by one or more Mach numbers, as the abort rocket would do. The pilots finally persuaded the program managers to find a way to terminate booster thrust rather than use the abort rocket. Terminating booster thrust was not an easy thing to do on solid rocket boosters, but they finally found an acceptable solution.

That concluded my two busy years of 1962 and 1963. I had flown fourteen different aircraft, including the T-33, T-37, L-19, C-47, F-101, several versions of the F-104, the Aero Commander, Schweizer 136 sailplane, the Paresev, M2-F1, and X-15. I had also lost my chance to go into orbit in the Dyna-Soar. I hoped the pace would decrease and I would be allowed to concentrate on one or, at most, two flight research programs.

3
THE PARESEV PARAGLIDER RESEARCH VEHICLE

After working on Dyna-Soar, there was no doubt in my mind that lifting reentry was the way to return from space. There was never a real concern in my mind about the ability of those lifting-entry vehicles to maneuver hypersonically or land horizontally. There was an aerodynamic heating penalty associated with lifting entry because the aircraft decelerated at a slower rate and was thus exposed to high heating rates for longer periods of time. But it seemed to me, that problem could be solved. The big hurdle was to get someone to commit to a lifting-entry configuration to demonstrate that it was a practical way to return from space, and that it offered the advantages of a benign g environment during entry and a precise landing on a preselected runway. I swore that I would accept nothing less in any program that I got involved with in the future.

I quickly had to renege on that oath when I got involved with the Rogallo wing paraglider. While I was working on Dyna-Soar, I happened to attend a lecture by Francis M. Rogallo, an engineer at Langley. He had been working since the mid-forties on the design of a wing-shaped parachute that could be steered to allow a precise landing. The Rogallo wing was intriguing to me. It was a manned kite—something almost anyone could build in the back yard. As a kid, I had jumped off a few roofs with large umbrellas, with sheets tied to my wrists and ankles, and with a few other unusual devices in an attempt to fly for a few feet. Nothing worked. The Rogallo wing was the poor man's airplane that I had been looking for as a kid. It did not have the glamour of a lifting-entry spacecraft, but it was an alternative to a parachute. It would enable a spacecraft to land on land rather than plunk into the water.

My primary reason for getting involved in the development of a manned paraglider was to demonstrate the landing capability of such a system. The Gemini and Apollo spacecraft configurations were already defined. I could not persuade anyone to change either spacecraft to a lifting-entry spacecraft. I could, however, possibly persuade someone to land on land rather than in the water. There was an advocacy group that was promoting a Rogallo wing for the recovery of the Gemini spacecraft. This group was not openly advocating a similar system for Apollo, but it may have done so if the Gemini worked as they envisioned. I supported the advocacy group when we began our manned paraglider. I wanted to help them demonstrate a horizontal landing on land using a Rogallo wing.

EVENTS LEADING UP TO PARESEV

Rogallo had persuaded NASA managers to take a hard look at his Rogallo wing for use as a spacecraft recovery system that would potentially enable the capsules to be recovered on land. Theoretically, his wing would allow the astronauts to maneuver the spacecraft to an airfield or a preselected landing site where it would glide down and land horizontally. In May 1961, Robert R. Gilruth, director of the Space Task Group, requested studies of a Rogallo wing for use on Gemini. An initial feasibility program was outlined that would enable competing contracts to demonstrate the potential of the concept. Three contractors were funded to develop concepts.

In November 1961, North American Aviation was selected the winner. North American received a contract to develop subscale Gemini capsules with deployable Rogallo wings for flight tests. These initial feasibility tests involved air drops of the subscale models from a helicopter to demonstrate deployment, inflation, and gliding flights of the wing and capsule. This was to be followed by the development, test, and demonstration of a full-scale paraglider for a Gemini capsule. The Rogallo wing for the Gemini was to be a stowable wing, equivalent in packed volume to a standard parachute. An inflatable wing structure was used to achieve the desired wing shape.

On November 28 and 29, 1961, a meeting was held with the various participants in the paraglider. The Dryden engineers left the meeting with some serious doubts about the paraglider. They believed that such a vehicle could pose a greater challenge than other advanced aircraft. They thought NASA should develop some experience before committing to an actual spacecraft entry. I talked with Paul Bikle about building a Rogallo wing and flying it in a low-cost, short-term program using part-time personnel. Bikle passed on the idea, as Dryden was committed to the X-15.

I then talked to Neil Armstrong about building one ourselves at home. Armstrong was interested, and we came up with a design and began scrounging parts to build it. Learning of our efforts, Bikle relented and approved an official program. I think he did this to prevent Armstrong and me from killing ourselves with our marginal design. The Paresev program (short for Paraglider Research Vehicle) was directed by Charles Richard, with Richard Klein, Vic Horton, Garry Layton, and Joe Wilson. Bikle told them to build a single-seat Paresev and to "do it quickly and cheaply." This was shortly before the 1961 Christmas holidays.

When Francis Rogallo was convinced that we were serious about building and flying a paraglider, he told us about an individual in North Carolina who had constructed and flown a paraglider successfully. He suggested we contact this individual and make arrangements to examine it and possibly fly it. Within a week, Vic Horton and I were on an airplane heading for the East Coast to meet Thomas Purcell, Jr., of Raleigh, North Carolina. Purcell was a friendly and outgoing person. He was enthusiastic about his paraglider and eager to let me fly it. The vehicle was somewhat disappointing to look at. It looked like a big kite with a seat suspended below it. It was definitely not an airplane. In fact, nothing about it looked like an airplane. The wing was constructed of three aluminum tubes with a cross brace between the center tube and the two outboard tubes. A clear plastic material was used as the sail between the tubes. Three small wagon wheels served as the landing gear. Overall, it appeared to be a rather flimsy structure. The most Mickey Mouse component was the pilot's seat: a standard folding lawn chair. That really destroyed our macho test-pilot image. I didn't want any pictures of me in that machine to get back to Edwards.

The vehicle flew surprisingly well. Purcell towed me down a small airstrip with his station wagon. I lifted off and flew down the length of the runway at a height of about 10 feet. We made several runs up and down the runway without releasing the towline. Purcell had never flown the vehicle in free flight, so he recommended that I stay on the towrope. Horton and I departed that same day to return to Edwards and begin construction of our own paraglider.

Horton and I also went to Ryan Aircraft Company in San Diego. The company had constructed a vehicle using a flexible Rogallo wing, and it was powered by a small aircraft engine with a propeller. They had flown it successfully and were attempting to sell it to the U.S. Army. Named Fleep (a contraction of "flying jeep"), it had not flown power off, so we still had no indication of what to expect in a free-flight glider. That was our primary interest in the Rogallo wing: its flight characteristics when used as a gliding recovery system. We still had to build our own vehicle to obtain that answer.

THE FIRST AIR TOW OF THE PARESEV

The Paresev 1 was finished in seven weeks at a cost of $4,280. The design engineers had used only simple drawings and sketches during construction. The framework for the Paresev was actually laid out with chalk lines on the floor of the welding shop and built without any formal blueprints. The framework was steel tubing, and it resembled a large tricycle, with the Rogallo wing perched on a tall tripod mast. The pilot (me) sat in a rudimentary seat with no enclosure of any kind. The wing pivoted on the mast and was controlled using an overhead control stick. The descent rate was controlled by moving the wing up or down, and the vehicle was turned by tilting the wing left or right.

It was the first NASA research vehicle built entirely in-house. Rather than conducting normal stress analysis, we subjected the Paresev to severe proof testing. We dropped the Paresev from a height of 42 inches to demonstrate its structural integrity in a 15-foot-per-second vertical velocity, 6-g landing. Although we weren't required to abide by most FAA (Federal Aviation Administration) regulations, Bikle intentionally sought FAA certification for the Paresev before we flew it. We applied for and received official aircraft registration numbers for both the Paresev and the M2-F1 wooden lifting body, which legitimated the airworthiness of those two vehicles before we flew them. I pitied the poor FAA representatives who had to certify the airworthiness of those two vehicles. We couldn't certify their airworthiness ourselves, even though we had conceived and developed them. We had to fly them to verify their airworthiness, regardless of all our scientific knowledge.

I made the first ground tow of the Paresev 1 on January 25, 1962. The Paresev flight program started with ground vehicle tows. An International Harvester Carryall was used as the tow vehicle. We ran up and down the taxiway behind the Dryden building to get a feel for handling the Paresev. The initial tests were done without lifting off, so I could check the control rigging and become familiar with the Paresev's ground stability. As I became more familiar with its behavior, the tow speed was increased, until I lifted off at about 40 knots. The Paresev remained attached to the towline. We next tried landing flares from an altitude of 10 or 20 feet using a slack towline; when I gave a signal, the Carryall driver would slow abruptly. I made several successful landings with this technique, and we decided it was time to release the towline and make a free flight. Here we began to run into trouble. Flying at 10 to 20 feet, there wasn't sufficient time to go from towed flight to free flight to landing. In 2 or 3 seconds, I had to correct for any transients due to the towline release, push over to minimize deceleration, and then perform the flare. We decided that a release altitude below 100 feet was undesirable.

The Paresev 1 was difficult to fly—I consider it more demanding than the later lifting bodies. The control system had a built-in lag, which made the Paresev 1 feel as if it were controlled by a wet noodle. The amount of control input varied according to the airspeed, and the control forces were high. Even at higher landing speeds of 60 knots, there was still only 3 to 3.5 seconds to initiate the flare. We were also disappointed to learn that the lift-to-drag ratio was 3.5 to 1 at 42 knots, rather than the predicted 4.2 to 1. There were two reasons for this. The wing was covered with Irish linen, which had been selected because it was commonly used on fabric-covered aircraft. The sailmaker who sewed the covering suggested Dacron instead. During flight tests we found that the wing flapped and bulged considerably. Added drag was caused by considerably more structure than the original drag analysis had allowed for.

After about two months of ground-tow flights, Bikle finally asked me when I was going to be ready for an air-towed flight to high altitude. His question created a momentary panic in my mind. My initial response to myself was, not for a long time yet. In fact, I felt insulted that he had even asked the question. I thought to myself, he actually believes that I am obligated to fly this ungodly machine up to an altitude where I might possibly get hurt if something goes wrong; I don't remember agreeing to do that. I also couldn't come up with any good reason to continue with ground tows. I had evaluated every possible flying characteristic that I could think of. While on ground tows, I had climbed up to altitudes of 100 feet or more, high enough to kill myself but still mentally acceptable because I was being towed by a ground vehicle and I could descend while on tow anytime I so desired. An air tow was a big step in my mind. Once I got airborne behind the tow plane, I was committed to go to altitude. I lost the option to land while still on tow. This rationale doesn't really make much sense, but that's why I hadn't been in any hurry to commit to an air tow.

There was a real added hazard in an air tow due to the wake of the tow plane. The wake of the tow plane included some strong vortices emanating from the wingtips of the aircraft. Those vortices could upset an aircraft that flew into them. The Paresev was not a very maneuverable vehicle. If it was upset or rolled over by the wake, it might be unrecoverable. I planned to stay well clear of the tow plane wake when I finally did make an air tow.

In response to Bikle's question, I finally said, "I guess I am ready." Deep down I knew that I had to bite the bullet and go. That is what I was getting paid to do.

Once the decision was made to do an air tow, we moved pretty fast to prepare for it. We acquired some small battery-powered radios to communicate between the Paresev and the tow plane. We procured a special parachute that was designed for rapid deployment and opening at low speeds. We also acquired some cold-weather flight gear, since it was still winter and I was completely exposed in the Paresev.

Paul Bikle contacted Fred Harris at Tehachapi to supply a tow plane. We scheduled our first attempt in early March 1962, but the weather didn't cooperate. I wanted perfect weather for the first flight: zero wind and no turbulence up to 5,000 feet, which was the intended release altitude. March isn't the best time of the year in the Mojave Desert for good weather. Generally it is cold and windy, in fact, very windy. We canceled because of weather on our second attempt a couple of days later.

Finally on March 12, 1962, we managed to get airborne. Harris was flying the tow plane, a Piper Super Cub. We used a 1,000-foot nylon towline to allow me to maneuver within a reasonable envelope behind the tow plane to achieve a safe tow position. Prior to the flight, we conducted a detailed briefing with all the flight and ground participants to cover the planned operation and any foreseeable emergency conditions. I made a strong plea to Harris to maintain a steady airspeed, within 5 knots, since we were not certain what the effects of airspeed were on overall flying qualities. That was something we hoped to learn during the air-tow flight program. We didn't necessarily want to learn that on the first flight, though. I wanted a benign flight with no surprises. The proposed ground track for the flight was a wide circle over the northwestern portion of the north lake bed. We wanted to stay well within gliding distance of the lake bed in case the towrope broke or some other emergency occurred that required a premature release from the towline.

During the briefing, I told Harris to hold his airplane on the ground until I got airborne and up above his wake. His wake was relatively weak prior to his becoming airborne. I lifted off the Paresev at roughly 40 knots and climbed up to a position about 50 feet above his airplane. Everything seemed to be satisfactory. However, the control forces in pitch were higher than I would have liked. Shortly after I got airborne, Harris began a slow turn to stay on our planned ground track. Within a few seconds, my airspeed began to decrease. I immediately called Harris and told him to watch his airspeed and maintain our planned airspeed as closely as possible.

Whenever the airspeed changed, I had to make adjustments in my controls to maintain the desired position above and behind his airplane. The radio reception was poor for both of us, so I ended up shouting at Harris almost continuously to gripe about his airspeed or flight-path control. In retrospect, we had created an unstable towing situation primarily as a result of using such a long towline and also because the rope was nylon. Whenever the tow plane turned, the airspeed of the Paresev tended to decrease, since the towline wanted to cut across the circle rather than following the path of the tow plane. To complicate the problem, there was a substantial amount of stretch in that 1,000-foot towline under load. The overall result was an unstable towing condition.

The airspeed during this process was varying continuously. Not realizing what the real problem was, I was blaming Harris and yelling at him for not maintaining airspeed precisely. All this maneuvering to maintain position quickly became tiring. The control forces were relatively high, and I was overcontrolling to maintain a precise position. My arms began to tire. The climb to release altitude was painfully slow. The Super Cub did not have a lot of excess horsepower.

It was a good tow plane for gliders or sailplanes with their very low drag, but it was overloaded with the 200 pounds of drag created by the Paresev. It took almost half an hour to reach the 5,000-foot release altitude, which was only 2,700 feet above the lake bed. I was almost exhausted when we arrived over the release point. I could only pray that the control forces would decrease after release. I finally released the towline and found that the control forces did decrease. If they hadn't, I thought I might have to bail out.

I rolled into a left turn to line up with Runway 18 on the north lake bed and then rolled out to make a constant-speed descent to a landing straight ahead. I maneuvered a little bit during the descent to evaluate the vehicle response, but my primary attention focused on positioning myself for the flare-out to landing. I wanted to have a steady-state glide at exactly 55 knots prior to initiating the flare for landing. The sink rate at 55 knots was roughly 30 feet per second, so I had to make a reasonably good flare to prevent damage to the vehicle or injury to myself.

I managed to make an excellent flare and touch down at less than 3 feet-per-second vertical velocity. Judging the proper time to flare was the major problem, since the flare maneuver didn't provide much leeway. I had a maximum of 4 seconds to complete the flare before running out of airspeed and stalling. The flight path prior to flare was negative 30 degrees, very steep compared with conventional airplanes. The flare maneuver was a major challenge, and yet each of the pilots who eventually flew the Paresev managed to make good landings during air-tow flights. They did have problems trying to flare from non-steady-state conditions during ground tow.

Once I stopped on the lake bed, everyone converged on me and the Paresev. Everyone was congratulating me and commending me on the flight. There was a real celebration and a general sense of relief that everything had occurred as planned. Bikle was particularly happy. In fact, he suggested that I make another flight. I wasn't at all receptive to his suggestion, since it had been a long and tiring flight; I was exhausted, both physically and mentally, and my arms were burning from the exertion. I had survived a traumatic experience, and I was ready to give up for the day and relax. I tried to convey that message to Bikle, but he was euphoric. Our homebuilt research vehicle had flown successfully. From the ground, the flight had appeared to be a huge success. After release the Paresev seemed to float down gently to the lake bed and land softly at the intended location.

After a few minutes my physical and mental condition improved significantly, and soon I was caught up in the excitement of making another flight to demonstrate that we had a successful flying machine. The ground crew hooked up the towline, and Harris fired up the engine on his Super Cub. An instant after I got airborne, I knew I had made a horrible mistake. The stick forces were still high and my muscles were fatigued. I knew I wouldn't be able to hang on to that stick for another half an hour until we got to the release point.

I considered releasing the towline early, but I was approaching the edge of the lake bed and I didn't want to land in the sagebrush. I finally wrapped my leg around the stick and controlled the vehicle with my leg and my hands. This time the climb seemed even slower. My arms were shaking with fatigue when we finally got to the release point. Again the control forces decreased to a manageable level after release, and I made it to the ground with my leg still wrapped around the stick to assist in controlling the vehicle. The landing was not quite as smooth as the first one, but I didn't bounce. This time I directed the ground crew to hook up the vehicle and head for the hangar. I wasn't about to be talked into another flight.

That afternoon the engineers and mechanics made some adjustments to the wing pivot point to reduce the control forces, and the next morning we were out on the lake bed bright and early to fly again. Before we made an air tow, I made a ground tow to check the control forces. They appeared to be much lower and acceptable for another air-tow attempt. Frank Fedor, the crew chief, hooked up the towline to the Super Cub, and Fred Harris and I took off for our third flight. This time things went much smoother. The control forces were much lighter, my anxiety level was much lower, and my confidence was greatly improved. We made two flights to altitude that morning, and I began to open up the flight envelope cautiously while increasing the amount of maneuvering. Both flights were very successful. We had an acceptable paraglider research vehicle. We could now begin exploring its useable flight envelope and demonstrate its potential as a spacecraft recovery system.

BRUCE PETERSON

That afternoon, Bikle suggested that I should check out Bruce Peterson, our newest pilot, in the Paresev so I could devote my time to more important projects. I was heavily involved in Dyna-Soar at the time, participating in the development of the Gemini parawing as a consultant, supporting X-15 flights, and supporting Dale Reed in his advocacy of construction of a lifting-body research vehicle.

I had to admit I was busy, but I didn't believe it was the proper time to check out a new pilot in the Paresev. We didn't know much about the Paresev, even though it appeared to be a simple vehicle to fly. Also, Peterson was a brand-new research pilot. He had transferred to the pilots' office only a week before, and he hadn't been exposed to any unconventional flight vehicles with strange or unpredictable characteristics. But Bikle persisted, and I, as usual, gave in to his strong pressure to concede. To dissuade Bikle, you had to have a strong and compelling argument. I didn't have one, so the next morning we towed the Paresev out on the lake bed and prepared to make some ground tows with Peterson at the controls.

I talked at great length with Peterson, describing the handling characteristics of the vehicle and what he should do during the ground tows. I advised him to stay close to the ground and remain in trail with the tow vehicle, particularly on his first flight. He could be a little more aggressive on subsequent flights, but I suggested he make the first one conservatively. Vic Horton (the project engineer), Bikle, I, and the driver climbed into the tow car and began to stretch out the towline. I called Peterson on our small radios to verify that he was ready to go, and then we started up the lake bed toward North Base.

As we accelerated through 35 miles per hour, Peterson rotated the nosewheel of the Paresev off the ground and shortly thereafter rose up slowly to about 15 feet above the lake bed in a slight bank to the left, then slowly banked to the right and abruptly nose-dived into the ground. Horton immediately released the towline, and the driver swung the vehicle around to head back to the Paresev. The flight lasted less than 10 seconds, and yet it appeared to progress in slow motion.

Everyone was astounded by the results. There was no obvious explanation for the vehicle's motions, and yet it seemed to fly directly into the ground in that bank reversal. When we arrived back at the vehicle within 30 seconds, the dust was still hanging in the air. The vehicle was almost completely demolished. The wing had broken off, the nose landing gear was gone, and numerous pieces were scattered about the impact point. Peterson's speed at impact was less than 50 miles per hour; however, the vehicle was obviously not designed to survive that kind of impact load.

The pilot's seat and the overturn structure did remain intact, but Peterson had been injured. He was in obvious pain, lying on his side in his seat, and blood was dripping from his gloved hand, which he was holding tightly to his stomach. The long control stick had broken into several pieces, and he had apparently been struck hard by one of the pieces. He was holding his hand over the place where the stick had hit his abdomen, and he was afraid to look under his hand. He thought a piece of the stick might still be embedded in his stomach. We immediately called on the radio for an ambulance and then attempted to make Peterson as comfortable as possible.

Peterson's first comment after we arrived was, "What happened to the stick forces?" Obviously they were excessive.

The ambulance took forever to get out to the lake bed. It was still early in the morning, and the duty ambulance hadn't left the base hospital for the flight line, so it had to drive almost ten miles to get to the crash site.

During this long wait, a photographer came out and photographed the accident scene. It was not intentional or obvious to those of us at the accident scene, but one photograph shows everyone standing off to one side calmly chatting and smoking cigarettes without paying any attention to Peterson, who was still lying in the debris. We had done everything we could to help him and then just waited for the ambulance. The picture appears to show some very callous observers.

Peterson was rushed to the hospital and thoroughly examined. The blood flow was from a cut on the hand that he was holding over the injured portion of his stomach. Something had hit him hard in the stomach, but it hadn't penetrated. He had several other cuts and bruises, but he was released from the hospital the next morning.

Peterson's entry into the flight research business was impressive, to say the least. Our secretary, Della Mae Bowling, summed up his first research flight succinctly in her daily log book: "Bruce fell approximately ten feet in Paraglider. Considerable damage to Paraglider. Bruce taken to Base Hospital but injuries do not seem to be serious."

THE PARESEV 1-A

Following the accident, we decided to build a completely new Paresev and try to correct some of the first vehicle's undesirable features. I particularly disliked the overhead control stick. I was also concerned about the attachment structure that joined the wing to the lower support structure. The redesign eliminated both of those concerns. We modified the control system to use cables to move the wing. These cables also served as a backup attachment of the wing to the supporting structure. To minimize fabrication time, we used as much off-the-shelf hardware as possible. The shock absorbers were Ford automotive parts, the wing universal joint was a 1948 Pontiac part, and the tires and wheels were Cessna 175 parts. We also took the opportunity to change the wing covering to Dacron, as the sailmaker had recommended. Subsequent ground-tow tests demonstrated that the wing would provide adequate lift. We estimated that the lift-to-drag ratio for the new wing would be 3.1; we subsequently found it was better than predicted: 3.9 at 42 knots.

The new vehicle, which we called the Paresev 1-A, was a big improvement over the original vehicle. The wing pivot was beefed up and was relocated to decrease the control forces. This vehicle looked as if it was airworthy. We soon learned it was, and also that it was easy to fly.

While the Paresev was being rebuilt, I was continuing to participate in the Dyna-Soar program and also supporting the X-15 program as a chase pilot, range check pilot, and lake-bed check pilot. On May 18, almost exactly two months after the Paresev crash, I flew the first air-towed flight of the new Paresev 1-A. It flew beautifully compared with the original vehicle. During a flight to obtain some publicity film, I could be seen flying hands off the control stick. I made a total of nine flights in the vehicle to altitude before we made another attempt to check out Peterson.

I continued to fly the Paresev occasionally, and in late June I began flying a smaller wing on the vehicle. No one else flew the smaller wing, which we called the Paresev 1B. The initial wing had an area of 150 square feet; the small wing had only 100 square feet. We wanted to determine the effects of wing loading on the vehicle's flight characteristics.

The original wing loading was 4.5 pounds per square foot, and the smaller wing loading was 6 pounds per square foot. The vehicle weighed roughly 600 pounds, which included the weight of the pilot and his personal equipment. Bikle decided that I should make the first flight of the small wing since I had the most experience. The only significant difference from the larger wing was the higher speed required for takeoff and landing and the higher speed required to make a successful flare and landing.

During this period we decided to do our own air towing, since it would be more convenient and cheaper. It turned out to be neither. It was, in fact, a bad decision, which almost cost the life of a pilot—me. An Army pilot assigned to NASA arranged for us to borrow an L-19 observing plane to use as a tow plane. It was an informal arrangement wherein we borrowed the aircraft during the week from an Army Reserve unit in the Los Angeles area, used it as a tow plane, and then returned it on Friday to be used for Reserve training on the weekend. We did not give the Army any details on how we were using the airplane.

Every Monday we would pick up the airplane and fly it to Edwards. We would then install the tow hook assembly, use it throughout the week, remove the tow hook, and fly it back on Friday. None of our pilots was checked out in the airplane, but I had flown the civilian version during my crop-dusting career, so I became the instructor pilot to check out the new tow pilots.

One of the first pilots that I checked out was Neil Armstrong. It was a quick and simple checkout because we wanted to fly the Paresev with the small wing the following day. Armstrong had never flown an L-19 before, nor had he flown anything as a tow pilot. I had to explain the towing operation to Armstrong, since we had no way of giving him experience before the test operation the next morning.

I explained the overall procedure and our normal ground track over the north lake bed. I emphasized the need to stay close to the lake bed so I could glide back and land there in case the towline broke or some other problem required a premature release. Also, I stressed the need for the tow pilot to make gentle turns to ensure that I did not lose airspeed during the turns. Early the next morning, June 28, 1962, the project team assembled out on the lake bed with the L-19 and the Paresev in preparation for the air-tow operation.

I again briefed Armstrong on the desired ground track to maintain during the climb to altitude and the need to maintain a constant airspeed during the climb. Our planned takeoff was to be made just west of and parallel to Runway 05-23 on the north lake bed. All of the north lake bed to the west of that runway was allocated to us by the tower for our Paresev operation. As soon as the L-19 and the Paresev were positioned for takeoff, the towline was attached. A quick check of the release mechanism was made in the L-19 and the Paresev. We were ready to go.

Armstrong taxied the L-19 forward until the towline was taut, and on my command over the radio, he applied power for takeoff. As we rolled across the lake bed, we were followed by a gaggle of chase cars and motorcycles, each raising a rooster tail of dust. The takeoff went smoothly. I lifted off in the Paresev as we accelerated through 45 knots and established a trail position approximately 50 feet above the L-19. The weather was good and the wind was calm.

As we approached the north shore of the lake bed, I advised Armstrong to begin his turn to the left to parallel the shoreline back to the west. He established what appeared to be a shallow turn, and I relaxed a little, assuming that Armstrong would do exactly what I asked him to do, since he had done everything right so far. I soon began losing airspeed as the turn continued. I could see a slight drop in the long, 1,000-foot towline, so I told Armstrong to "speed it up." The radio communications were not too readable, so I repeated my request, assuming he had allowed the airspeed to decrease during the turn.

My request apparently did not get through to him, since my airspeed continued to decrease. Our altitude at the time we initiated the turn was not high, less than 500 feet above the ground. As I began losing airspeed, I also began losing altitude. I couldn't maintain altitude at the lower airspeed without an undesirable excessive increase in angle of attack. I called again and told Armstrong to speed it up. Unknown to me at the time, Armstrong was assuming that I wanted him to speed up the *turn*. He dutifully increased his angle of bank each time I asked him to speed it up.

The results of this miscommunication were temporarily funny but ultimately almost disastrous. In my last visual image of the operation, Armstrong was abeam of me heading south in the L-19. I was headed north in the Paresev, and the towline was hanging in between us. That's one of the disadvantages of using a 1,000-foot towline. I couldn't believe what I was seeing, and I was still losing airspeed and altitude.

I finally realized my only recourse was to release the towline and attempt an immediate landing straight ahead. I had one new major problem, however. I was outside the lake bed shoreline and over the sand dunes that were covered with sagebrush and mesquite. I couldn't get back to the lake bed. All I could do was grit my teeth and hope for the best.

I made a good flare and came level just above the sagebrush, but I was traveling about 50 knots. I hit the ground, bounced back into the air a few feet, and then slammed into the ground nose first. It was a tremendous impact, but I didn't get hurt. I waited until the dust settled a little before releasing my seat belt and shoulder straps. I was hanging just off the ground. The second my feet touched the ground, I began running away from the vehicle, afraid it might begin burning. After a couple of steps, I asked myself why I was running. The vehicle had no fuel on board. It was just an ingrained response from my early flight training.

As I stood there in the sagebrush, I thought, "Now some damn sidewinder will bite me and I'll die of a snake bite." While I was feeling sorry for myself, Armstrong continued along the planned ground track, which passed over Dryden at its southernmost point.

Armstrong was not sure what had happened to me. I had not had time to call him before I released. The crew on the ground saw the L-19 pass overhead with the towline still attached but nothing on the end of it. What a shock. Armstrong immediately landed, and after some discussion they sent out several vehicles in a search party to locate me and the Paresev. They finally found me about half an hour later, sitting on the wreckage smoking a cigarette.

Four weeks later, we were flying again. I made several more flights with the small wing, and then we reinstalled the larger wing to begin gathering data. There was little wind tunnel data on the Rogallo wing and no wind tunnel data on our specific vehicle configuration.

One of the more important measurements of interest to us was the lift-to-drag ratio. We had no instrumentation system on the Paresev, so we had to make do with a mixture of data from pilot readouts and ground tracking cameras. The ground cameras were precision tracking cameras that defined the vehicle's position in space accurately.

We installed an angle-of-attack vane with a scale at the apex of the wing. On the pilot's instrument panel we had an altimeter and an airspeed indicator. To obtain data we would tow the Paresev up to altitude and then position it in the field of view of the ground cameras. The pilot would then release the towline and establish a constant airspeed. During the descent, he would call out airspeed, angle of attack, and altitude at increments of 500 feet. By matching the photos of the descent with the pilot's data reports, we could plot curves of lift to drag versus airspeed and angle of attack.

This procedure worked reasonably well and provided good data. However, on occasion the ground camera operator had trouble locating the Paresev through the lens. Someone had the bright idea of installing a smoke bomb on the Paresev to make it more visible. After release from the tow plane, I was to activate the smoke bomb and then set up a constant-speed glide. On our first and only attempt to use the smoke bomb, the smoke spewed out of the bomb but, instead of trailing behind the Paresev, began circulating around under the wing.

I was immediately blinded and was quickly gasping for air. That acrid smoke was burning my eyes, nostrils, and throat. I could barely see my instrument panel. I was indeed flying blind, and there was no way to shut off the smoke bomb. For the next 30 seconds, I hacked and coughed and tried to wave the smoke away while I rapidly approached the ground. The observers on the ground told me later that the Paresev looked like a ball of orange smoke with a wing on top and some wheels sticking out the bottom. Thankfully, the smoke burned out before I hit the ground. I never did forgive the engineer who came up with that idea.

We had a number of other interesting incidents during this flight phase. On one flight, just before release from the tow plane, the needle of the airspeed indicator came off the shaft and fell to the bottom of the instrument. What a disaster. There was no safe way to fly the Paresev without an airspeed indicator. I finally decided to remain on the towline and have the tow plane descend so that I could land while still on the towline.

On another occasion, we replaced the nylon rope in the rear seam of the wing with a cable to minimize the fluttering of the sailcloth wing membrane. As I lifted off in the Paresev the next morning, the vehicle began to roll to the right. We apparently had an asymmetry in the wing's trailing-edge lines from the center spar to the outer spars. I managed to release the towline and level the vehicle before I hit the ground. The lateral stick forces were horrendous.

This incident vividly demonstrated that we were groping in the dark in our attempts to fly the Paresev. We had no good wind tunnel database and no way to predict the flight characteristics of the vehicle through analysis. A pilot was killed in a subsequent accident in a different vehicle when he inadvertently decreased the wing angle of attack excessively. The longitudinal stick forces recorded during the recovery attempt exceeded 200 pounds. These two incidents diminished my enthusiasm for paragliders. They could be so deceivingly graceful to fly as long as they stayed within a limited flight envelope. If the pilot inadvertently exceeded those limits, the graceful soaring eagle bared its cruel talons.

We made another stupid decision in an attempt to save a little money. The towline that we had been using was half-inch nylon rope. It had some undesirable characteristics when used for towing, but we had adapted to them. However, we were using a lot of rope. Our policy in regard to rope damage was conservative. We discarded the rope after ten flights, or if it was visibly damaged from being dragged along the lake bed. Those 1,000-foot towlines were relatively expensive for our limited budget.

Vic Horton had the bright idea of using steel tow-target cable to save money. It wouldn't scuff or chafe, and better yet, it was free. There were used cables scattered all over the base and the lake bed. All we had to do was find them, pull them out of the mud or the sagebrush, and check them for visible damage. I wasn't too impressed with the idea, but I didn't have any good reason not to give it a try. Most of the cables had rusted to some extent after lying out in the weather, but we pull tested a few of them, and they seemed to have retained most of their strength.

As soon as I got airborne, it was obvious that the cable had some disadvantages. I had to use a lot more force to keep the Paresev in position above the tow plane. That cable weighed quite a bit more than the rope, and I had to fly at a higher angle of attack to lift the extra weight.

As we began maneuvering during the climb to stay within our test area, another undesirable characteristic of the cable became apparent. It had a pronounced pendulum effect, swinging from side to side as we maneuvered. It was a nervous ride to altitude on the end of that damn cable. I was relieved when I finally released the towline and began my glide back to the lake bed. I promised myself that I would wrap that cable around Horton's neck once I was safely on the ground.

In late September 1962 we began a series of checkouts, or demonstration flights, in the Paresev. Neil Armstrong was the first pilot to be checked out. He had just been selected as an astronaut, and we believed this was a good opportunity to demonstrate the paraglider capability to an astronaut who might eventually fly one in a Gemini capsule.

I checked him out just before he transferred to Houston. We had notified Houston before we flew him to determine whether they had any objections. We didn't want to risk sending them a damaged astronaut without their permission. The checkout went well. We spent two days ground towing and then towed him up to altitude for a couple of free flights. On his last flight, he made a spot landing next to our ramp onto the lake bed and rolled out heading up the ramp.

Gus Grissom came out in late October to check out in the Paresev. I flew him up to Tehachapi to check him out in a sailplane for some towing experience, then we came back to Edwards to begin ground tows in the Paresev. The checkout sequence that I had established involved ten to fifteen runs up and down the north lake bed flying 10 to 20 feet above the lake bed while on tow. During these runs, the pilot would maneuver gently behind the tow vehicle to feel out the control system and vehicle response. Once he felt comfortable, he would release the towline to evaluate any transient motions and then land.

The second series included flights up to 200 feet or more above the lake bed. On reaching 200 feet, the pilot would release the towline, push over to achieve an airspeed of 55 knots, and then flare and land. The timing of the sequence of events after release was quite critical. The pilot had only a few seconds to push over, gain the necessary airspeed, and then initiate the landing flare. Grissom failed to get everything done in the allotted time. He hit hard and broke the nose gear off the vehicle. While we were collecting the pieces off the lake bed, Grissom proceeded to chew me out for planning such a critical maneuver. He, like Bikle, believed that we should progress from low-altitude flights directly to air-tow flights during checkout.

Air tows provided much more time after towline release to set up for the landing. I had to agree, but my twisted logic told me that was an awfully big step. I still believed deep down that one should creep up to altitude in small steps, even though I knew I could get killed once I exceeded 20 or 30 feet of height. I finally did change my mind though. During one of my own flights to check out a modification to the Paresev, I got caught in a bad situation while attempting to make a flight similar to the one that Grissom made when he broke the nose gear.

I was being towed by the Carryall van. It didn't have a lot of excess horsepower. As I attempted to climb to 200 feet, the Carryall couldn't maintain speed, and as a result I would slow down and lose altitude. I tried several times as we proceeded up the lake bed, but each time the same thing happened. On the last attempt we were rapidly approaching the north end of the lake bed. I still couldn't get to 200 feet, but I knew the tow vehicle would have to stop before reaching the edge of the lake bed, so I released the towline and pushed over to gain airspeed. I didn't make it.

I didn't have enough altitude to gain the necessary speed to make the flare. I made a partial flare but ran out of altitude and energy. I hit hard. The nose gear broke off, and I skidded to an ignominious stop. That was the last time that we tried a free flight from that altitude. All future checkouts involved only the low-altitude flights on tow before making an air tow. I finally had to admit that I was wrong, but not before we had chalked up three broken nose gears.

The crew chief eventually painted the names of each pilot who broke a nose gear on the frame of the vehicle. The Paresev has been displayed in the Experimental Aircraft Association Museum in Oshkosh, Wisconsin, and more recently in the Smithsonian Annex at Dulles Airport. I have not seen it since it left Edwards, so I don't know if the names are still there. I assume so.

Grissom stayed at Edwards until the Paresev was repaired and then made several air tows. He asked me how high we had flown the Paresev. I told him 6,000 feet above the lake bed. He said he would like to set a record. On his next flight we towed him up to 7,000 feet above the lake bed. He made a beautiful flight.

In early November, I checked out Bob Champine, a NASA Langley research pilot and a veteran of NACA and NASA research. He had flown the first NACA X-1 research aircraft in the late forties, along with a number of other exotic research aircraft. During the early Mercury program he spent a lot of time on the centrifuge at Johnsonville demonstrating the human ability to tolerate extreme g levels. This was done to support the development of the escape rocket on the Mercury capsule. That rocket was eventually designed to accelerate the Mercury capsule off its booster at 13 g. That's a lot of g's.

The last two pilots to be checked out in the Paresev were Emil Kleuver, an Army pilot attached to Dryden, and Charles Hetzel, a North American Aviation test pilot who was scheduled to fly the test flights of a full-scale Gemini capsule suspended under an inflatable Rogallo wing.

INTRODUCTION TO SAILPLANES

It was in the spring of 1963 that Paul Bikle decided that I should get a commercial glider license. I had been flying the Paresev for more than a year and had been checking out various pilots in the vehicle for several months, many of whom had never flown a glider or sailplane. I had been reasonably successful, but occasionally a pilot would have a problem during towing operations. Bikle decided that I should check out each new pilot in a glider prior to the checkout in the Paresev. The towing operation was the most hazardous part of any glider operation, and Bikle felt it would be safer to check out new Paresev pilots in a standard glider before exposing them to the unconventional Paresev.

Bikle and I flew up to Tehachapi the following week in our Aero Commander to do some glider flying. He put me in the front seat of a Schweizer 232, and he climbed in the back. The tow pilot hooked us up and, after starting his own airplane, began the takeoff. Bikle indicated that I should do all the flying and he would kibitz. I must have done a pretty fair job of flying on tow, because I got no adverse comments from Bikle, who was a tough taskmaster and quick to criticize. He asked me to change tow position several times, and then we finally released and glided back to the field for a landing. During the descent, he asked me how I had learned to fly on tow. I told him I had no previous instruction. Whatever I was doing that was correct had just come naturally. I did indicate that formation flying probably had a lot to do with my success. Bikle was particularly intrigued with my technique of maneuvering around behind the tow plane while maintaining a wings-level attitude. He indicated that most new glider pilots attempted to fly from one position to another rather than translate in a wings-level attitude. Many pilots created hazardous situations by attempting to fly back and forth behind the tow plane. Some were killed, particularly if the situation developed at low altitude.

Bikle sent me out solo on the next flight. I was really impressed by a number of things on that flight. It was so quiet and smooth: no engine noise, no significant aerodynamic noise, no radio noise since there was no radio, no equipment noise such as electrical motors whirring or whining, and no vibration or shaking. I heard a train. I rolled over and sure enough, a train was passing beneath me. I was 4,000 feet above the train, but I could still hear it clearly. I heard an air horn on a big cement truck below me on the highway. During the descent to the field, I reached the low key position with a couple of hundred feet of excess altitude. I executed a 360-degree turn to kill off some altitude and still ended up about 50 feet high. A turn like that in the Paresev would have required almost 1,000 feet of altitude to accomplish. During the landing approach, I was attempting to make a spot landing such that I could roll up to the parking area off the midpoint of the runway. I did quite well. The spoilers were quite effective in dissipating excess energy. I stopped right in front of Bikle and the tow pilot, who were standing in the parking area.

Bikle decided that I should have the best instructor locally available to qualify me for a commercial glider license. Gus Briegleb was his choice. I flew Bikle and Walter Whiteside over to El Mirage Dry Lake in our C-47 for my first official glider lesson. Briegleb was extremely thorough. He briefed me on all aspects of glider flying and then put me in the front seat of one of his TG-3 glider trainers. From the time we started rolling down the runway on tow until we touched down to a stop, Briegleb was talking. His talking varied from instructions to commands to compliments to cursing to cajoling. I felt like I was back in Navy flight school on my first training flight. Bikle and Whitey said they could hear him hollering at me as we flew over them on the lake bed. After we landed and climbed out of the glider, Briegleb was still bending my ear about the flight. To qualify me for a commercial license, he required a minimum of ten hours of instruction. I wasn't sure that I could survive nine more hours of that kind of instruction. I didn't complain, but Bikle decided that I probably didn't need that degree of training.

Bikle and I went back to Tehachapi for my next training session. Fred Harris, the owner-operator of Tehachapi, had been our primary Paresev tow pilot. He was somewhat more pragmatic about the training requirement. He rationalized that if I could fly the highly unconventional Paresev, I should qualify for a commercial sailplane license.

To make it official, he instructed me to make a solo flight around the pattern after being towed up to 1,500 feet and to land and stop within 10 feet of a line painted across the midpoint of the runway. The flight lasted 10 minutes. In another 10 minutes I had Harris's signature on an application for a commercial glider license. I would now be legally qualified according to FAA regulations to check out other Paresev pilots.

NASA did not really require that I have such a license. At that time we did not have to abide by either FAA or military regulations. As NASA pilots we could sign our own flight plans or, if flying VFR (visual flight rules), not even file a flight plan. The Air Force actually issued a special regulation, AFR55-26, exempting NASA pilots from all Air Force regulations in the conduct of aeronautical and space research. We didn't flaunt this exemption often, but when we needed to violate some prescribed rule or regulation, we sometimes quoted the authorizing regulation, and occasionally we got into shouting matches with operations duty officers at military bases away from Edwards.

PETERSON'S CHECKOUT IN THE NEW PARESEV

Following his crash in the Paresev 1, Bruce Peterson had made ground tows in the rebuilt Paresev 1-A, flown the L-19 for several air tows, and, on August 27, 1962, entered Class 62C of the USAF Test Pilot School. Two of his classmates were Capt. David Scott, who later landed on the Moon on Apollo 15, and Capt. Michael J. Adams, who was selected for the X-15 program and was later killed in the crash of X-15 number 3.

By late January 1963, I was overloaded, trying to fly the Paresev, consulting on the design of the Gemini parawing, participating in the Dyna-Soar program as a pilot astronaut, and preparing to fly the M2-F1 lifting body. Something had to give.

The decision was made to turn the Paresev program over to Peterson. By this time, we had successfully flown the rebuilt Paresev over a period of eight months and made more than fifty air-tow flights. We had obtained a lot of data and had successfully checked out three other pilots (Neil Armstrong, Gus Grissom, and Bob Champine).

We had accomplished a lot, but there was much more to do to support the development of the Gemini parawing. One of the first issues to be addressed was how to get the Gemini capsule and parawing up to altitude so that it could be tested in free flight. The proposed Gemini capsule test program would begin with the wing deployed and inflated, raising the question of how to lift the Gemini with the wing deployed. One proposal was to tow it up to altitude with a helicopter, but this posed a potential controllability problem for the Gemini capsule pilot, since flying on tow was not a straightforward control task. Another proposal was to lift the combination of capsule and inflated wing and then drop them, hoping they would glide properly without tangling the suspension and control lines.

Houston and North American needed some information on these issues. We offered to investigate the options with our Paresev. Thus began some of the most dangerous tests in the program. Peterson worked with the U.S. Army helicopter unit at Edwards to obtain a helicopter capable of lifting or towing the

Gemini capsule. He then began a series of tests towing the Paresev behind the helicopter to determine the feasibility of that procedure. The major concern was the rotor wash. It could easily cause a complete loss of control in the Paresev. Peterson suggested marking the rotor wash with smoke so that he could see its position during the towing operation. It worked quite well, and he made a number of successful air tows behind the helicopter. We never did demonstrate the other option—dropping the Paresev—since we could not satisfactorily simulate the wing and capsule drop problem.

Peterson did address a more basic question with the Paresev: Would the inflatable wing actually develop lift? The inflatable wing did not look like a real wing. To me, it resembled a triangular life raft. The inflatable booms that served as the wing structure were quite large compared with the size of the wing. Those booms were the wing's leading edges. They seemed to be much too ungainly to serve as leading edges. The best way to find out if the wing would develop lift was to build one and fly it.

North American built a half-scale wing, which was a reasonable size to test using the Paresev. We had to modify the Paresev to attach the wing and its control cables, but this was not a major job. The only obvious problem was that the inflatable wing weighed considerably more than our Paresev wing, which would affect the controllability of the Paresev. In fact, it would affect it much more than we initially anticipated.

We called the modified vehicle the Paresev 1C. Peterson's first attempt to get airborne during a ground tow of the inflated wing was made on March 14, 1963. Just like his first attempt to fly the original Paresev, it ended with a crash. The Paresev lifted off, rolled slightly to the left, then began rolling to the right, and nosed into the lake bed. The control forces were excessive, particularly in roll. The Paresev was subsequently modified during its reconstruction to attach the rudder pedals to the roll control cables. This enabled the pilot to use his legs to assist in moving the wings for roll control.

Peterson made a successful flight in the Paresev 1C on March 26, 1963. Subsequent ground- and air-tow tests demonstrated that the inflatable wing would provide adequate lift.

GEMINI PARAGLIDER TESTS

While the Paresev program had achieved a measure of success, the Gemini paraglider tests, which were going on at the same time, had run into problems.

North American's initial feasibility tests involved air drops of the subscale models, followed by tow tests (behind a helicopter) of the model capsules, which would then be released and glide to a radio-controlled landing. This would be followed by manned tow tests of a full-sized capsule behind a helicopter.

The director of the Manned Spaceflight Center requested Dryden assistance in conducting the tests. Vic Horton was selected as the project engineer, and I was selected as the flight operations representative to conduct the tests. With helicopter support from the U.S. Army test unit at Edwards, we proceeded to drop the subscale models over the Rosamond Dry Lake. This was my first operational experience with helicopters. The lift helicopters used for these tests were fairly heavy for that time. The main rotor was driven by two R-2800 Pratt and Whitney radial engines, connected to the main rotor through massive gear boxes that were inside the cabin. I often wondered what would happen if I tossed a handful of bolts into those two gear boxes. I could picture an explosion of gear teeth riddling the fuselage of the helicopter. That made me concerned about something accidentally getting into those gear boxes. They were exposed and vulnerable. I did not like what I saw. Helicopters were dangerous. I was thankful that I wasn't a helicopter pilot.

The subscale vehicle tests during 1962 and 1963 were marginally successful. The parawing was able to establish a stable glide, but there were repeated difficulties in deploying and inflating the parawing. That was the real problem—when something looks so simple in construction people are deceived into thinking that it must be simple to fly. This was an instance where that was not true.

NASA made a final attempt to demonstrate an operational parawing recovery system for the Gemini capsule. North American was given a contract to build and fly an improved suspension and control system for the Gemini parawing. The full-scale simulated Gemini spacecraft was to be the testbed for the flight tests, and it was to be piloted for these tests. The plan was to lift the inflated parawing and capsule by helicopter up to altitude and then drop it to achieve a free-flight glide to a landing.

The helicopter lift operation worked quite well. Charles Hetzel, who had been checked out in the Paresev, made the first captive tow flight on July 29, 1964. The helicopter towed him aloft, they flew around the area for about 20 minutes, then Hetzel landed still attached to the towline. On August 7, the first attempt at a free flight ended in disaster. When the Gemini was released, it immediately went into a tight left-hand spin. Hetzel bailed out of the uncontrollable capsule, breaking a rib in the process. The Gemini suffered minor damage on impact.

Over the next five months, North American made fourteen radio-controlled flights using subscale models, followed by two unmanned tests of the full-scale vehicle on December 15 and 17.

The next manned test of the Gemini parawing was made on December 19, 1964, by Donald F. McCusker. The release and 5-minute glide were successful, but the landing was extremely hard. McCusker misjudged the landing flare and hit the lake bed at a vertical velocity in excess of 30 feet per second. That was a hard landing! In fact, it may not even qualify as a controlled crash. McCusker was injured by the shock. It was not easy to judge when to flare a paraglider. Judging flare initiation in the Paresev was not straightforward, and it did not have the control-system lag inherent in a cable suspension and control system.

North American attempted to beef up the Gemini landing gear and pilot seat to withstand vertical velocities of 35 to 40 feet per second, but there was no simple way of protecting the pilot at those extreme touchdown conditions. McCusker and other North American test pilots made twelve more Gemini parawing flights between September and November 1965 before NASA finally canceled the program. All the Gemini flights used a standard parachute and water recovery. It was unfortunate that a satisfactory parawing system could not be developed in time for use on the Gemini capsule. It would have greatly simplified the overall recovery operation if the Gemini could have been landed on land.

Francis Rogallo also proposed a hypersonic Rogallo wing that would be capable of returning spacecraft from orbit or might serve as an astronaut lifeboat, but no one picked up on the idea. In the early sixties a hypersonic Rogallo wing did not appear to be practical. Neil Armstrong at one time facetiously talked about a space dirigible—at least I thought he was being facetious. We did not possess the materials that would survive the extreme heating associated with hypersonic flight. We now have some fabric materials that are capable of withstanding temperatures of 1,200 degrees F.

Just a few years ago, in discussions about the space station, the possible need for an astronaut rescue vehicle surfaced, a vehicle that would serve as a lifeboat if a major problem occurred on the station. I began thinking about Rogallo's rescue vehicle concept but imagined instead an inflatable lifting body made of these new high-temperature fabrics. I really became enthused and did some simulation to evaluate the concept. One intriguing result of the simulation was that the inflatable lifting body with a very low wing loading (less than 5 pounds per square foot) would decelerate from orbital speed to subsonic speeds while still above 100,000 feet altitude. This meant that the heating rate would be extremely low even though the stagnation temperatures were high. Not much heat would be transferred to the lifting body with so few molecules of air impacting it. An inflatable dirigible or lifting body just might work.

SUMMING UP THE PARESEV

The Paresev made a total of 341 flights in just over two years. The last air tow of the vehicle, on August 6, 1963, was piloted by Bruce Peterson. The final ground tow, using the Paresev 1C wing, was also flown by Peterson, on April 14, 1964. Seven pilots flew the vehicle: Bruce Peterson, Neil Armstrong, Gus Grissom, Robert Champine, Emil Kleuver, Charles Hetzel, and myself. Between us, we broke the nose gear six times. Tow vehicles ranged from the Carryall, a Super Cub, a Cessna L-19, and a Stearman biplane to a Boeing HC-1 helicopter.

The Paresev and the Gemini parawing programs showed it was not a simple task to develop a good, reliable parawing system. Various forms of parawings, parafoils, and other gliding parachutes have evolved over the last thirty years. Some of them have excellent performance and handling characteristics. There have been some impressive demonstrations by sport jumpers and military parachute jumpers, but there have been problems developing reliable systems for large payloads. Various military organizations have attempted to develop parawing systems to deliver cargo to precise landing locations with limited success. NASA has made numerous attempts to develop parawing systems for various cargoes. There has been some measure of success, but not without some operational problems.

I know from personal experience of four or five cases where competent, experienced pilots have had serious accidents with paragliders, hang gliders, or ultralights. In many cases they have serious stability or control problems. Yet, they are easy to fly and relatively safe when flown within a limited flight envelope. Surprisingly, inexperienced pilots can sometimes adapt to these vehicles better than the best test pilot. I personally would not fly them or recommend that someone else fly them; I'll leave that to the more adventurous.

I'll stick with rocket planes.

Ultimately I did lose interest in these systems for use as spacecraft recovery systems. They are much more complex than a simple parachute, and they have limited energy management capabilities. In strong winds, the vehicle will land going backward. The Russian parachute-landing rocket system appears to be the most practical system for land recovery of a capsule-type spacecraft. I would prefer a lifting-entry spacecraft that could be landed horizontally, such as the lifting bodies or the shuttle orbiter. That is the most practical and efficient spacecraft recovery concept developed so far.

4
M2-F1 EVOLUTION

I was sitting at my desk in the pilots' office when I heard footsteps pounding down the hallway. It was Dale Reed towing a small model on the end of a long string. He was demonstrating the flying qualities of a lifting body during towed flight. Reed had been advocating the construction and flight test of a manned lifting body for several weeks, and now he was taking this opportunity to demonstrate the feasibility of his proposal to several senior engineering managers, who were lining the hallway. The demonstration was impressive: the model was very stable on tow.

Reed had proposed that the manned lifting body be towed rather than self-propelled to simplify the vehicle. It was to be a lifting-body glider. A towing demonstration enhanced the practicality of his overall proposal. A few minutes later, Reed was up on the roof of the building launching the model in free flight to glide out to the aircraft ramp behind the building. Again, the model flew surprisingly well. The senior engineering managers were impressed, but Paul Bikle, the director, wasn't.

Bikle had a much more challenging vehicle to test—the X-15. It was February 1962. The X-15 flight program was progressing on schedule, but problems were being encountered and additional resources were required to keep the program on schedule. Bikle did not have any excess manpower to devote to a quixotic flight into the future. Lifting bodies were potential spacecraft configurations. Bikle had to demonstrate that we could fly safely at Mach 6. He was in no hurry to tackle a flight to orbital speeds just yet.

Reed persisted. On the weekends, he attempted to tow the lifting-body model into the air with a conventional model aircraft. He managed to succeed after several attempts and came back on a Monday with some 8-millimeter movies of the two models in flight. Bikle was impressed but resolute in his decision to deny Reed the resources to proceed to develop a manned vehicle. Reed realized he needed some strong support, since the chief of the Research Division, Tommy Toll, was also skcptical about supporting thc proposcd program. Toll was more enamored with the Rogallo wing as a means of flying a capsule to a landing. We were then in the early tow tests of the Paresev, which were to demonstrate the maneuverability and horizontal landing capability of a simple Rogallo wing paraglider.

Reed knew Bikle listened to the pilots. Bikle had worked and flown with test pilots throughout his entire career. He respected their opinions largely because he knew they were exposed to the broad picture in a test program. They had to be on speaking terms with the aerodynamicist, the propulsion specialist, the structures engineer, the stability and control expert, and the flutter analyst or structural dynamicist, at a minimum. More bluntly, their butts were on the line if something went wrong. Bikle tilted the scale in their favor should an impasse occur. He also valued their opinions on new programs. Bikle was somewhat biased since he, too, was a pilot, not a professional test pilot, but a competitive sailplane pilot who held a world altitude record of 46,000 feet in an unmodified competition sailplane.

Reed moved into the pilots' office with his models, sketches, aerodynamic data plots, and home movies. He rubbed snake oil on his models and interspersed seductive mirages in his movies. I was vulnerable to his mystical sales pitch since the Dyna-Soar program was beginning to founder. I could sense that I would never fly back from space in a Dyna-Soar spacecraft. I was extremely disappointed that we would not demonstrate the feasibility of a lifting-entry and horizontal-landing spacecraft. The lifting-body concept was an attractive alternative to the winged Dyna-Soar. It offered the advantages of volumetric efficiency (as everything was contained within a wingless fuselage) and aerodynamic simplicity (as a lifting body was a simpler shape than a winged spacecraft).

The program that Dale Reed was proposing was by no means a substitute for the Dyna-Soar program. It was several orders of magnitude smaller in scope and hundreds of millions of dollars cheaper, but it was a start. We would worry about how to get it into orbit later. First we had to get it off the drawing board. Reed's proposal involved the construction of three different lightweight lifting body configurations: a blunt cone shape with a slightly contoured base called the M-1L, a slender half-cone shape with a contoured base called the M-2, and a lenticular shape. Each of these shapes had some form of stabilizing and control surfaces.

Under Reed's plan, one basic internal structure and three different outer shells in the aforementioned shapes were to be constructed. The internal structure was to be constructed of steel tubing. It would include the landing gear, a major portion of the flight control system components, the tow hook, the pilot's instrument panel, the pilot's seat, and the overturn structure. Each of the three outer shells would include the pilot's canopy and the stabilizing and control surfaces. The internal structure and the three shells were to be constructed using materials and techniques commonly used in gliders and sailplanes. The vehicle was to be towed into the air by a powered aircraft and then released and flown in free flight as a glider. During the course of the test program, the outer shell would be changed to test each of the three shapes.

The program was intriguing and innovative—I liked it. I wrote a letter to Paul Bikle supporting it. I also talked to him. I recommended that we begin by building only one outer shell, the long slender cone M-2 shape. At least that shape looked like it might fly. The M-1L blunt cone was too radical, and the lenticular shape was too exotic, though there were many reported sightings of alien flying saucers. Somewhat surprisingly, each of these shapes would in reality develop lift and thus fly, although not very efficiently.

Almost any shape will develop some lift traveling through the air. One of the few shapes that won't is a sphere or ball, but almost anything else will. As a demonstration, a few of the old aerodynamicists would mold a piece of clay into a ball and then test it in the wind tunnel to prove that it would not generate any lift. Then they would throw it against the wind tunnel wall and test it again. Lo and behold, the flattened ball would generate some lift.

BIRTH OF THE LIFTING-BODY CONCEPT

Lifting-body configurations were a spin-off of research being done to develop missile nose cones that would survive the intense aerodynamic heating of atmospheric reentry. Ballistic missiles are boosted out of the atmosphere in a ballistic trajectory to a target several thousand miles away. The missile's speed determines the distance that the missile and its nose cone will travel. The nominal speeds for an intercontinental missile are 16,000 to 18,000 feet per second or 11,000 to 13,000 miles per hour. When the missile nose cone reenters the atmosphere on the way down to the target, the friction of the air heats up the nose cone to temperatures in excess of 3,000 degrees F. That temperature will melt most metals. This intense heating would also melt and distort the tip of the nose cone, causing both aerodynamic and structural load problems. The sharper the tip, the more severe the melting problem. The aesthetically pleasing, slender needle-nose shape would not survive a reentry through Earth's atmosphere, so the missile aerodynamicists began looking for help in designing shapes that would.

A team of NACA aerodynamicists, H. Julian Allen, Alfred Eggers, Clarence Syvertson, George Edwards, and George Kenyon, at the Ames Research Center began looking at other shapes. Blunt nose cones were tested, and they looked promising. The team also began rounding the tip of the cone. They observed that as the cone shape became blunter and the nose more rounded, a strong aerodynamic shock wave developed in front of the nose cone. This strong shock wave drastically reduced the heating on the nose cone, decreasing the maximum temperature to levels compatible with existing metals and nonmetallic materials. Further investigation revealed that the nose roundness, or radius, was the most significant factor in minimizing the heating problem. The nose cone could be relatively slender as long as the nose was well rounded. The ballistic missile nose cone survivability problem appeared to be solved. Intercontinental ballistic missiles were feasible.

Allen, Eggers, and the other NACA aerodynamicists noted during their developmental testing that these nose cone shapes developed lift when they were misaligned with the wind stream. They didn't develop much lift, since they were clearly not a classical lifting shape like a wing airfoil, but they did provide enough lift potentially to influence the flight path of the nose cone during reentry. This was interesting. It implied that a vehicle might actually fly at these extreme speeds and altitudes. Theory predicted this, but no one had devoted much effort to verifying the possibility. No one had thought much about flying at these speeds except the Germans who were developing and flying the V-2 rockets during World War II. They conceived an intercontinental bomber that would be boosted up to hypersonic speeds and would then skip along the top of the atmosphere before descending to glide to its target and blow it up. They proposed to bomb New York using this concept.

The early rocket designers also noted this. As Al Draper, a noted Air Force aerodynamicist used to say, they referred to lift as the "positive increment of drag." Those early rocket scientists and engineers were not enamored with the science of aerodynamics. In fact, they would prefer that there was no atmosphere. The atmosphere just created problems for them. It created drag, stability and control, structural, and aerodynamic heating problems. They didn't need oxygen from the air to run their rocket engines. Their job would be greatly simplified if the atmosphere evaporated.

The NACA engineers were intrigued by the possibility that the lifting blunt nose cone concept could be used to build a spacecraft that could be flown back from orbit by a human pilot. If such a "lifting body" could be built, it would open a range of possibilities. Even with a lift-to-drag ratio as low as 1.5 to 1, the g forces of reentry could be reduced from the 8 g of a conventional nose cone to 1 g—normal Earth gravity. The vehicle could also maneuver during reentry and make a precise landing anywhere within a sizable footprint. They began looking at additional modifications to the basic cone shape that would increase the lift and thus the lift-to-drag ratio or aerodynamic efficiency. They set out to develop a shape that would fly at hypersonic speeds and be capable of a horizontal landing like an airplane. Those two objectives are not necessarily compatible. Anything that flies well at low speed and can land like an airplane is not the optimum shape to fly at hypersonic speeds. Conversely, anything that is shaped to survive the extreme heating at hypersonic speeds cannot readily be landed. Wingless cone shapes were not known to fly well at any speed. Their normal aerodynamic application was for such things as bombs, artillery shells, and bullets. Rather than landing softly on Earth at the conclusion of a flight, they were intended to create monstrous holes or inflict death.

In early 1957, several months before *Sputnik* was launched, Eggers had developed the first true lifting-body shape. The M-1 was a half-cone, flat on the top, with a rounded nose and flaps around the base for steering. Its hypersonic lift-to-drag ratio was 0.5. The following year it was refined into the M-2 shape, which had a better ratio of 1.4 to 1 at hypersonic speeds. At subsonic speeds, however, it was prone to tumbling end over end. Eggers and the other Ames engineers boat-tailed the top and bottom of the aft end, giving it more of an airfoil shape. The final version added a canopy and two fins, earning it the nickname "M-2 Cadillac." The requirement to be able to fly and land at subsonic speeds dictated the final configurations of these lifting bodies, much more so than the high-speed flight requirements. A simple cone without any stabilizing fins or streamlining modifications would fly adequately at hypersonic and supersonic speeds. It could bank, turn, pull up, or push over as long as the pilot had control either from aerodynamic surfaces or reaction control thrusters. At subsonic speeds more aerodynamic sophistication was required.

During this same period, aerodynamicists at the Langley Research Center also developed an interest in lifting-entry spacecraft. They began testing a series of shapes oriented toward a lifting-entry spacecraft rather than a ballistic missile nose cone. Their shapes were a much more radical departure from the simple symmetrical shapes that the missileers started with. The HL-10 (for "Horizontal Landing") configuration that they ultimately developed was cone-shaped in one axis only. In the other axes, it resembled an airfoil shape. The result was a modified delta shape. The Air Force was also paying attention to lifting bodies and was starting development of the SV 5 shape, which resembled a finned potato and was later flown as the X-24A. The M-1 and lenticular lifting-body shapes were contenders during the selection of the shape for the Apollo capsule. They were not selected primarily because they were relatively complex aerodynamically. The objective of the Apollo program was to put a man on the Moon before the Soviets did. That dictated that the vehicle be as simple and light as possible.

Thus, by February 1962, there was a whole family of possible lifting-body shapes. Despite the wind tunnel data indicating excellent hypersonic aerodynamics, many doubted their stability at supersonic or subsonic speeds. Reed had been following the questions about their low-speed stability and realized Dryden was in a unique position to answer those questions. He felt that if a lifting body could demonstrate a horizontal landing, it would build confidence in the concept throughout NASA, and the advantages would be quickly realized. Reed contacted Eggers, who was now the Ames deputy director, and proposed the idea of building a manned lifting body. Eggers supported the idea and told Reed to pursue it. This he did with the single-minded determination of the maverick innovator, even in the face of derision by some engineers of his hallway model test.

Reed, a Dryden engineer named Dick Eldredge, and I prepared a plan to build the wooden lifting body. We persuaded Eggers to come down from Ames to hear our presentation to Bikle. Eggers enthusiastically offered wind tunnel support for the project. Bikle approved a six-month feasibility study. By the fall of 1962, the feasibility study was complete, and Bikle gave the go-ahead to build the lightweight M2-F1 manned lifting body. Officially it was a "full-scale wind tunnel model." Bikle noted that if the aircraft "just happened" to be capable of flight, that was something that was out of management's control.

M2-F1 CONSTRUCTION

Once Bikle approved construction of the lifting body, he assumed the role of program manager. He decided that we would build the internal structure but would contract out the construction of the outer shell. Work on the M2-F1 began in September 1962. Bikle had a section of our calibration hangar curtained off as a construction area. A sign was put up reading "Wright's Bicycle Shop." The curtains were intended to provide a low level of security and, more important, to minimize the number of sidewalk superintendents. Bikle preferred to maintain a low program profile during vehicle construction. He did not want the program questioned or canceled before it had acquired some momentum. Bikle ran the program out of local discretionary funds and used the volunteer help of the aircraft homebuilders at Dryden.

Our mechanics, sheet-metal smiths, and technicians began cutting and welding metal tubing the day after the curtains were hung. The basic shape of the internal framework became evident within the first weeks of construction. The primary engineering tool for the internal structure was a model made of wooden doweling. Some parts, such as the wheels and brakes, were purchased off the shelf. The main wheels used on the M2-F1 were Cessna 150 wheels. The main gear shock used a cylinder with a loosely fitted piston and 50-weight motor oil. As with the Paresev, we did drop tests to ensure we had the desired degree of damping.

As I had suggested, the only outer shell built was the M-2 shape. Bikle's decision to have a contractor build it was based on Dryden's lack of experience with wood construction and the problems we had had with the Paresev sail. He called a few of his friends who were involved in sailplane construction and asked if they were interested in building the outer shell for a unique glider. Gus Briegleb, a local sailplane builder, indicated that he was interested. Briegleb was the designer and builder of the TG-3 trainer sailplane. Bikle showed him sketches of the lifting body and the two major structural components, the steel-tube internal structure and the outer hull.

Both Bikle and Briegleb decided that the hull should be constructed of plywood to ensure that the body contours were as smooth as possible. The NASA aerodynamicists were not sure what the effects of surface irregularities might be on the flying qualities of the vehicle, but they unanimously agreed that smoother was better. Standard wooden glider construction materials and techniques were used. The hull had a 3/32-inch mahogany plywood skin and 1/8-inch mahogany rib section reinforced with spruce. The exterior was wrapped with Dacron and doped for a smooth, durable surface.

Bikle asked Briegleb what he might charge to build the hull. Briegleb, after due consideration, estimated that he might have to charge as much as $10,000. He was somewhat defensive in justifying his estimate, since $10,000 was a lot of money in those days. You could buy a complete new sailplane for less than that. He was therefore surprised when Bikle seemed to think he was underestimating the job. They argued for a while, but Briegleb wouldn't change his estimate, so Bikle accepted his bid and had our procurement officer write out a contract. The entire procurement process was that simple. Briegleb made a bid and Bikle accepted it. If they tried to do that under current rules and regulations, they would end up in a federal prison. Even in those days, Bikle was breaking the rules, but there were no whistle-blowers; no one made an issue of minor procurement infractions. They were more interested in getting the job done.

When Briegleb finally finished the hull structure, he totaled up his costs to build it and found his estimate had been quite accurate. The cost of construction was indeed $10,000, but that did not include any profit. When Bikle became aware of this, he revised the contract to include some extra minor work and then added another $2,000 in payment for it. Briegleb had done a beautiful job building the hull. Bikle believed he deserved a profit.

While construction was under way, the major question in my mind was whether the lifting body had the performance required to make a successful landing, performance, that is, in terms of lift and drag to complete a landing flare. The early predictions of the lift-to-drag ratio based on model tests were encouraging, indicating a ratio of 5 to 1. Our previous experience with the X-15 and the Paresev demonstrated that we could make successful landings with vehicles with lift-to-drag ratios of 4 to 1 and 3 to 1, respectively. Thus, we should be able to land a lifting body with a ratio of 5 to 1.

The concern that I had was whether the M2-F1 flight vehicle would actually have a ratio of 5 to 1. The flight vehicle would be much dirtier aerodynamically than the small wind tunnel model that had produced the early performance data. The small-model results were obviously optimistic, but how optimistic? No one knew. The real flight vehicle would have irregularities, gaps, and discontinuities that the small models wouldn't duplicate. In addition, the flight vehicle would have fixed landing gear projecting from the fuselage. That would create a lot of drag, which would significantly reduce the lift-to-drag ratio.

In an attempt to resolve the question of whether the vehicle could be landed, we developed a simulator using wind tunnel data and began an extensive investigation to determine the best technique for consistently successful landings. We quickly established the need for a high-speed approach to ensure that we could complete the flare and then have a few seconds to adjust the rate of sink before we ran out of airspeed. In the simulation, we could vary the lift-to-drag ratio to evaluate the best landing technique for the most optimistic and pessimistic values. For the worst case, it appeared that we needed a minimum speed of 120 knots prior to flare to make consistently successful landings.

With that requirement established, I asked the engineers who designed the vehicle what the maximum design speed was. I was informed that it was 110 knots. I then informed them that I needed a minimum of 120 knots to make a successful landing. We appeared to have a major problem. The engineers went off into a huddle and then came back and said that the vehicle was good to 120 knots. I asked them how they decided that. They didn't have a good answer. That didn't give me a lot of confidence. They had been adamant about the maximum design speed of 110 knots, but when I challenged them, they rather nonchalantly increased the allowable speed to 120 knots.

I did know that they were being conservative in the structural design of the vehicle since it was such an unconventional configuration. I assumed that the extra 10 knots was well within the design margin and that's why they were willing to agree to the higher design speed.

BILL DANA TUMBLES IN THE M2-F1

The M2-F1's steel tube internal framework was completed several weeks before Briegleb delivered the outer shell. Dale Reed and Vic Horton decided that we should conduct some ground-towing tests to evaluate the high-speed stability of the landing gear system on and off tow, and the nosewheel steering system. We spent a day or so towing the internal structure up and down the lake bed, while I maneuvered it behind the tow vehicle. I wasn't too happy with the nosewheel steering system, so we decided to modify it. When they completed the modification, I was not available to check it out. I was in Seattle at Boeing participating in a Dyna-Soar suit evaluation. Vic Horton called me in Seattle to suggest that they proceed with the evaluation using Bill Dana as the evaluation pilot. I disagreed with his suggestion and strongly recommended that they wait to evaluate the new system until I returned.

On December 6, 1962, I got a call from Della Mae Bowling, the pilots' secretary, informing me that Bill Dana had been injured during a tow test of the M2-F1 internal structure. The project management had ignored my recommendation and asked Dana to check out the new steering system during some ground-towing tests.

The new system was a copy of the Cessna 150 nosewheel steering system, which used a heavy spring instead of a solid linkage between the rudder pedals and the nose gear to move the nosewheel. To me, that was a stupid steering concept. Its touted advantages in a conventional airplane were that it provided nosewheel damping and a relatively linear response to pedal input regardless of speed. This would minimize any sensitivity problems as speed changed. I had complained of steering sensitivity problems during the initial tests. This concept worked reasonably well in a conventional aircraft because the effect of the vertical tail was to increase aerodynamic directional stability. At higher speeds, nose gear deflection was less for any given rudder pedal deflection than it was at low speed.

The internal structure, however, had no aerodynamic stability, something the project engineers overlooked. During the towing tests with the new system, Dana did a lot of maneuvering behind the tow vehicle at various speeds and finally moved off the trail position as far as possible using the maximum rudder pedal deflection available. He wasn't getting much lateral displacement because the towline was naturally resisting any deviation from the trail position. Dana finally decided to release the towline from this position while he still had maximum rudder pedal deflection applied. When he released the towline, the nosewheel snapped to full deflection, since it no longer had any resistance from the towline or any aerodynamic stability from the vehicle itself. The internal structure immediately tumbled. The tumbling was quite violent, since the ground speed at release was in excess of 70 miles per hour. Dana suffered a dislocated shoulder and various bruises to his body and ego. The vehicle sustained only minor damage.

THE PONTIAC TOW CAR

By early 1963, the internal structure had been repaired, the outer shell was delivered, and we were making final preparations to fly the M2-F1. We did have one little problem. Before we made towed flights behind the C-47, Bikle decided, we would make ground-tow flights until we understood its handling qualities. We had done this with the Paresev, but the M2-F1 was a lot bigger, a lot heavier, and had to be towed a whole lot faster before it would fly than the Paresev had. The Carryall we had used to tow the Paresev wouldn't cut it.

Bikle and Reed decided a souped-up high-performance car would work as a tow vehicle. Bikle selected Walter Whiteside to find a suitable vehicle. Whitey was a real fix-it, go-get-it-done type. He was also one of a large contingent of hot-rod enthusiasts among the Dryden employees. Most of us lived in Palmdale or Lancaster, a good 40 or 50 miles away, and there was an abundance of long, straight roads that were lightly patrolled by the local police. Whitey got out his slide rule and calculated the speed and horsepower we would need. We wanted to test the M2-F1 in ground tow up to 104 knots, which was about 95 percent of the design speed. We would need the lightest possible car with the biggest engine.

Whitey finally selected a 1963 Pontiac Catalina convertible. General Motors provided a 421-cubic-inch, triple-carburetor Tripower engine like those being run at the Daytona 500 race. The car was also equipped with a four-speed transmission, a hot cam, and a heavy-duty suspension and cooling system. This was done without any publicity; the whole M2-F1 project was being done under the table, and the last thing we wanted was some accountant wondering why we needed a muscle car. A sedan would have been lighter, but we needed the convertible for a clear view of the M2-F1 during tow operations.

Whitey picked up the car on January 9, 1963, and took it to a body shop for modifications. The interior was gutted, a roll bar was added, radio equipment was installed, the right-side passenger seat was turned to face aft, the rear seat was removed, and a second seat was added for a side-facing observer. Once this was done, Whitey took the Pontiac to Mickey Thompson's shop in Long Beach on January 25. The engine was torn down and brought up to spec. Special headers were also added. The work was completed on January 29, and Whitey drove it to Edwards. He came in the back gate and drove the car directly into a NASA hangar. We gave it government plates, painted "National Aeronautics and Space Administration" on the doors, and covered the hood and trunk with a garish high-visibility yellow paint.

In February, Whitey and new NASA pilot Don Mallick made a series of airspeed calibration "flights" in the Pontiac. They grabbed their clipboards and helmets and drove out to Boron, California, and Highway 395. Out on the desert there were measured mile markers that allowed them to calibrate the car's speedometer up to its maximum of 120 miles per hour. There were also not very many state highway patrolmen out there. While this was going on, I was up in Seattle working on the Dyna-Soar, being checked out in the Aero Commander, and making a series of flights in the Douglas F5D.

By the end of February 1963, the M2-F1 was completed and checked out, and the Pontiac had completed its breaking-in period and was also ready. Dr. Hugh L. Dryden, for whom the center would be named, said that the purpose of flight research was to "separate the real from the imagined and make known the overlooked." We were about to make the first tow tests of the M2-F1. Over the next twelve years, until the lifting-body program ended in 1975, we would be doing just that. In the process, we would be laying the groundwork for the space shuttle.

5

THE M2-F1 ALOFT: LEARNING TO FLY WITHOUT WINGS

On March 1, 1963, we rolled the M2-F1 out of the hangar for the first attempt to get it airborne behind the Pontiac tow car. The crew hooked the M2-F1 to the 1,000-foot towline, which was attached to the Pontiac. We checked the tow hook release a couple of times to ensure that it operated correctly and then towed the M2-F1 out on the lake bed. We set up to make our tow runs on the west side of Runway 18. The lake bed in that area was quite smooth and generally free of debris.

Much of the lake bed was littered with old shell casings, ammunition belt links, live machine-gun and cannon rounds, tow target cables, and various other aircraft-related paraphernalia. The lake bed and surrounding area had been used as a bombing and gunnery range during World War II, and most of the debris on it was a result of that activity. All of the debris did constitute a hazard to aircraft operating on the lake bed, since it could cut tires or in some cases be sucked into engine inlets and damage the engine.

When the M2-F1 was in position, the crew in the Pontiac released the towline and made a run up the lake bed to check for debris and potholes. We had approximately 4 miles of lake bed available to the north of our ramp. This would provide ample room to accelerate up to speed and then maintain it for a minimum of 2 minutes to evaluate the handling characteristics of the M2-F1.

Our first couple of runs up and down the lake bed were made at low speed to allow me to evaluate the ground steering and braking characteristics. During these runs, I released the towline a couple of times to determine whether there were any transient motions induced by towline release. There were none. On successive runs, we began building up speed as I evaluated the aerodynamic control forces and their effectiveness while still on the ground. I managed to raise the nosewheel off the ground during these tests and found that I could control the nose position quite accurately. Based on those results, I was ready to try to get airborne.

On the next run, the Pontiac crew accelerated up to 80 knots and then stabilized at that speed. Simulation and analysis indicated that I should be able to lift off and fly at 75 knots. I gradually eased the nose up, and as predicted, the M2-F1 became airborne, bouncing from one main wheel to the other. It was uncontrollable in roll. I couldn't keep the wings, if you'll pardon the expression, level. We tried a couple more runs, but I still couldn't adequately control the vehicle in the roll axis. The vehicle just bounced from one main gear to the other. Dale Reed suggested we install the center fin and try again.

We towed the M2-F1 back to the hangar, installed the center fin, and made another run in an attempt to get airborne. The results were the same. We had major handling deficiencies. During our postflight debrief, we discussed the problem and any potential solutions. No one had any good suggestions on how to cure it. The simulator had not predicted this problem; however, we were not certain that the wind tunnel data used to develop the simulator was of the highest quality. None of the wind tunnel models accurately replicated the configuration of the flight vehicle.

An old saying goes, "Flying without feathers is not easy." We were learning that flying without wings wasn't either.

TESTS IN THE AMES WIND TUNNEL

In the wake of the failed attempts, we decided to take the M2-F1 up to the Ames Research Center in the San Francisco Bay area and test it in the full-scale 40-by-80-foot wind tunnel. We were eager to fly, but we realized we had to solve the controllability problem, and a wind tunnel test was the only promising solution. We notified the appropriate engineers at Ames that we wanted to test the M2-F1 and asked for priority in scheduling the tests. The M2 configuration had been developed at Ames, so we had some strong advocates to support our request. Within a week we had the vehicle in the wind tunnel.

A minor problem arose when we discussed the test procedures. During the tests, the control surfaces had to be moved to various positions to obtain control effectiveness data. We had no simple way of positioning the control surfaces during a test. Our solution was to put someone in the vehicle to position the controls and hold them steady during the tests. Since I was the pilot and the one who had the most to lose if we didn't get good data, I was volunteered to sit in the vehicle during the tests. Sitting in the vehicle during testing might not sound like such a big deal, but in reality it was almost as dangerous as flying it.

The vehicle was mounted on three struts that projected 20 feet above the floor of the tunnel, to the middle of the test section. The test section was rectangularly shaped with rounded corners. It was 40 feet high and 80 feet wide. The struts were instrumented to measure the forces generated by the air flowing around the test vehicle. I had to climb a 20-foot ladder to get into the cockpit and then settle down with the canopy locked for a long, lonely test series. The ladder was removed prior to the initiation of the tests, and the door into and out of the tunnel was sealed. I was on my own in that huge tunnel with only radio contact with the outside world. If the M2-F1 was not sturdy enough, it could come apart or break off the support struts and fall to the floor of the tunnel. Worse yet, the high-speed air flowing through the wind tunnel would blow the vehicle or the pieces into the turning vanes behind the test section or even into the huge propellers that accelerated the air in the tunnel up to high speed.

I remembered how casually the engineers had raised the maximum design speed of the vehicle to 120 knots, so it was unnerving to be sitting in the M2-F1 as the wind speed increased. I was listening carefully for any strange noises that might indicate an imminent structural failure. Of course, there was little I could do if I did hear strange noises. The airflow in the tunnel could not be stopped instantaneously. It took several seconds to slow those big propellers down. By that time, I might be mincemeat, having been blown through the turning vanes and into those propellers.

I spent many hours sitting in the M2-F1 in that wind tunnel, and I finally quit worrying about a structural failure. In fact, I began proposing a test that would enable me to fly the M2-F1 on a towline that would be attached upstream of the test section. I would be on a tether and would be able to lift off and fly up and down and side to side in the test section. The wind tunnel operators thought about my proposal but soon turned me down. They were afraid that I might have a control problem and slam into the walls, floor, or ceiling of the test section. If that happened, I and all the M2-F1 pieces would be blown down the tunnel and into the propellers. I tried to convince them that it was still safer for me than traveling behind the tow car at over 100 miles an hour. I couldn't convince them. I think they were more concerned about damaging their wind tunnel than they were about injuring me.

Between tests, the Dryden project personnel began a contest to see how far an individual could run up the side of the wind tunnel. The side walls of the test section were semicircular. By building up some speed on the floor of the tunnel, someone could climb 10 to 15 feet up the side wall. The problem was that the runner lost momentum as he climbed, along with the centrifugal force that held him to the surface of the side wall. If he was very agile, he could turn around as his vertical motion ceased and start running back down the side wall. The severity of the problem increased as the runner reached higher and higher positions on the side wall. He eventually would be standing almost horizontally on the side wall at the end of a high-speed run, in an unrecoverable position. He couldn't build up any centrifugal force on the way down; he would simply tumble down the side wall to the floor. The side walls and the floor were made of steel. It was painful to slam into the wall on the way down. I'm amazed that someone didn't break a neck or at least an arm. Instead, only painful bruises were incurred.

THE SECOND ATTEMPT TO FLY

The wind tunnel data from the tests of the M2-F1 revealed some major differences from the small-scale model tests. The lift-to-drag ratio of the flight vehicle was 3 to 1, substantially lower than the small-scale model results. I would definitely need the maximum allowable speed prior to flare to make a successful landing. The full-scale wind tunnel results also differed from the small-scale tests in control effectiveness and static stability. We modified our simulator to include the new data and then reran some handling-quality evaluations.

The results suggested a change in the configuration of the control system. The previous simulation indicated poor roll response from the elevons but good roll response from the rudders. Based on those results, I had recommended that the rudders be linked to the lateral stick motion and the elevons to the rudder pedal motion, a somewhat bizarre control-system configuration but one that might provide the desired results. This was the configuration implemented for the first flight attempt on March 1. It was, however, unflyable.

When I attempted to lift off, I induced a lateral oscillation that had the vehicle bouncing from one main gear to the other at a relatively high rate. Much of this was a result of my anxiety and overreaction to any motions that I perceived immediately after liftoff. My personal gain was too high. The basic problem, however, was that the rudder surfaces produced two different moments, a rolling moment and a yawing moment. The rudder rolling moment was the first perceived motion. The rudder yawing moment was the second perceived motion, and it ultimately produced an opposite rolling motion, due to the dihedral effect, which substantially overpowered the initial rudder rolling motion. This second rolling motion was the desired response. To the pilot, however, the initial roll response due to the rudder was in the wrong direction.

This created considerable confusion in my mind, particularly when I was operating in a high gain mode. When I tried to correct for the initial motion, I was stimulating a pilot-induced oscillation immediately upon liftoff. This control configuration was flyable up and away and on the simulator, but it was unflyable close to the ground. The simulator did not predict this problem, since we did not have a motion base simulator or a good visual system to simulate a high pilot control gain. I had made a poor decision on the control-system configuration.

The new full-scale wind tunnel data indicated better roll response from the elevons. I also reconsidered whether I needed a lot of roll power. I finally decided to change the control system to a more conventional configuration for the next flight attempt.

The wind tunnel tests on the M2-F1 were completed on March 15. On March 18, I left for Brooks Air Force Base to undergo the Dyna-Soar physical, which lasted through March 22. March 27 through April 4 were taken up with F-104 flights. Along with the time needed to analyze the data and make the modifications to the control system, this delayed our next flight attempt.

The second M2-F1 flight attempt was not made until April 5, 1963. I managed to get airborne after a couple of bounces and stay airborne for longer and longer periods of time. After a couple of runs up and down the lake bed, I could lift off and fly the entire distance, approximately 4 miles, without touching down inadvertently. I wasn't getting very high on the first few runs, generally staying within 10 feet of the lake bed surface. Whitey read off the Pontiac's ground speed, while I radioed my altitude. In a high-stress situation, I usually reverted to the old cliché of flying low and slow to avoid getting hurt. Needless to say, the most probable way that I could get hurt was to impact the ground, but that didn't persuade me to climb up away from the ground.

That first day with the new control configuration, we made eleven runs up and down the lake bed. It was late in the day when we finished the last run, and the wind was picking up a bit, so we returned to the hangar. It was a satisfying day. We had accomplished a lot. We demonstrated that the vehicle would fly at the predicted speeds and that it was controllable within a limited flight envelope. We now would have to expand that flight envelope.

PREPARING FOR AIR TOWS

During the spring and into the summer, I made a number of ground-tow flights in the M2-F1. We expanded the vehicle's speed envelope and made the first limited free flight to understand its handling. I would climb to about 200 feet, pull the nose up into the flare attitude, then release the towline and glide down to a landing. I also did two demonstration ground tows in the M2-F1 for Dr. Robert C. Seamans, the NASA associate administrator, on May 8. The ground-tow flights provided only a limited amount of data on the M2-F1's free-flight handling, however. We had to wait until we were ready to start making air tows, and that required several more months of work.

It became evident during our simulation studies that the flare and landing task would be challenging. The simulator predicted that I would have less than 10 seconds to complete the flare prior to touchdown under the best of circumstances. That was better than the 3 seconds that I had available in the Paresev that I had previously flown. The descent rate, however, was substantially higher in the M2-F1 than the Paresev: 4,700 feet per minute versus 2,400 feet per minute. Thus the task was more demanding. I had to reduce the rate of descent to 600 feet per minute or less at touchdown or risk breaking the landing gear. I could usually achieve acceptable touchdown vertical velocities on the simulator, but occasionally I would exceed the allowable 600 feet per minute. Our simulator was rather crude in terms of its visual display, which may have contributed to the inconsistent results, but the overall results were not encouraging. We needed something to improve our odds of making good landings consistently.

One solution that came to mind was a small motor to provide thrust to extend the time from flare initiation to touchdown. The simplest motor would be a solid rocket motor similar to the jato bottles that were used during World War II to provide additional thrust for heavyweight takeoffs. Jato rockets were available through the military, but they were much more powerful than we needed. We needed something in the 200- to 300-pound thrust range. A motor of that size that burned for 8 to 10 seconds would effectively double the time available from flare initiation to touchdown.

We searched through the government supply lists for a rocket motor of that size but came up empty-handed. Vic Horton had a friend who worked at China Lake, where the U.S. Navy developed many of its weapons. He told Horton that the Navy had nothing available in that thrust range but might be able to make something for us. They routinely mixed batches of solid rocket propellant for various weapons such as missiles. After mixing a batch, they constructed a small test motor casing, which they filled with the newly mixed propellant and then fired to determine the characteristics of that batch.

These batch motors were possible candidates for our application. Horton's friend indicated that he could build some special batch motors to our specifications if we were in no hurry to get them. This offer was even more appealing when he said he would not charge us for the motors. We couldn't possibly pass up that offer. We asked for ten motors with 250 pounds of thrust with a duration of 8 to 10 seconds. Horton's friend was true to his word. He produced the motors some four months after we made our request and invited Reed, Horton, and me up to China Lake to observe a test firing. We flew up on May 7, 1963, early on the morning of the test firing, and were given a tour of the various facilities at China Lake. It was an extremely interesting tour.

The Navy was conducting a wide variety of projects involving weapons. One that particularly intrigued me involved the fabrication of explosive rocks that could be used as land mines. Horton's friend pointed out a bunch of rocks in a small yard in front of the building where he worked. They were dummy rocks that the Navy had created to match the rocks found in various parts of the country. At that time they were also producing some rocks that matched those native to Vietnam. During the remainder of our visit, I avoided every rock that I saw. That project was right out of Agent 007 novels and movies. To me, it wasn't fair to make explosive rocks, regardless of the old cliché "All's fair in love and war." It just didn't seem sporting to fool an enemy like that.

The batch motors worked as advertised. We installed one in the M2-F1 and then fired it to verify that we had a good system. We also fired one during a ground-tow test to verify that the rocket was properly aligned through the vehicle's center of gravity, minimizing any potential stability or control problems. All of our tests were positive. We had a good emergency landing assist system, labeled Instant L/D, since the rocket effectively doubled the maximum lift-to-drag ratio of the M2-F1 when it was fired.

The only negative aspect of the rocket system was the cost. Several months after the rockets were delivered, we received a bill for $10,000 from China Lake. Our friend's superiors did not go along with the idea of giving the rockets to us for free. They totaled up all the material and labor costs and sent us the bill. We were shocked not only because they reneged but also because of the amount of money they wanted. We had constructed the complete vehicle for less than $20,000. It really pained us to have to pay half as much again for those small rockets. In those days it was common to barter goods and services among government agencies. China Lake turned out to be a pariah.

During the ground-tow flights, the M2-F1 had been equipped with a standard bucket-type seat. Prior to commencing air-tow flights, we planned to install an ejection seat in the M2-F1. We wanted a lightweight seat that had a zero-zero capability, meaning that a successful ejection could be made at zero airspeed and zero altitude. The plane could be stopped on the runway or taxiway, and the pilot could still eject safely. Even before the first ground tow, we reviewed the specifications for the various ejection seats that were in production and decided to use the seat from the T-37 aircraft, made by Weber Aircraft in Burbank, California. We made several trips to Burbank to tailor the seat to our needs by recommending several modifications. Part of the process involved weighing me in the seat to determine a center of gravity so we could properly align the nozzle on the seat rocket. Our modifications required a requalification of the seat after they were incorporated.

The requalification test involved firing the seat and a dummy out of a simulated M2-F1 cockpit, which was located down at the south shore of the Edwards lake bed, near the sled track. On the first test, the seat ejected properly and climbed on up to the planned altitude, but the dummy did not separate from the seat properly and the parachute failed to deploy. The seat and the dummy slammed into the desert together. The next test occurred several days later. On that test, the seat ejected properly and the dummy separated from the seat, but the parachute still failed to deploy. Again, the seat and the dummy slammed into the ground while the spectators cringed and covered their eyes. It's really a disturbing sight, to behold an unsuccessful ejection sequence. You know it's just a dummy, but you can't prevent yourself from reacting as though it were a human being hitting the ground. It's frightening, but it's also funny.

The next test was scheduled about a week later. The day of the test, we received a call that everything was ready and the test would commence as soon as we arrived at the south lake bed test area. Six of us piled into one of our Carryalls and roared down the ramp onto the lake bed. The test site was about

12 miles to the south of Dryden via the lake bed route. I was driving, and I put the pedal to the metal. We were traveling between 80 and 90 miles per hour, kicking up a dust cloud a couple of hundred feet high. About a mile beyond the Old South Base runway, I noticed a dark line up ahead that extended across my intended track. Normally I would have ignored it and kept the throttle wide open. This time I let up on the gas pedal and then started braking as the black line began to widen. I finally slid into a panic stop as the black line evolved into a huge crack in the lake bed surface. The crack was about 3 feet wide and 2.5 to 3 feet deep. It extended to the east and west as far as the eye could see. It was a sobering sight. We had never seen any crack of that magnitude in the lake bed surface. We finally found a way around the crack and proceeded on down to the test site, this time at a slower pace.

The test was a success. Everything worked perfectly. The dummy separated from the seat just before reaching the peak of the trajectory, and the parachute deployed and inflated properly. The dummy touched down softly under the chute, and everyone cheered. We were ready to go. The requalification was declared a success, even though we had only one successful test out of three attempts. We were in a hurry to fly the M2-F1 to altitude; we didn't want to wait for additional tests. Besides, subsequent tests may not have been successful. That would have been a serious blow to my confidence and our flight schedule. We didn't want that.

The new seat ultimately demonstrated a perfect success record in the Lunar Landing Training Vehicle (LLTV). Our seat system was also selected for use in the Lunar Landing Research Vehicle (LLRV), which we were flying in support of the Apollo program. When our development tests of the LLRV were completed, we transferred our two LLRVs to Houston. They and three additional LLTVs procured from Bell Aircraft were used to train the Apollo astronauts. During the next couple of years, three LLTVs were lost during training and checkout missions. In each case, the pilot or astronaut made a successful ejection from very low altitude and unusual attitude. I was particularly pleased, since I had originated the modifications to improve the overall operation of the seat and parachute.

The crack in the lake bed we found the day of the successful seat test was a harbinger of future problems. Following the test, we drove back up the lake bed toward Dryden. When we reached the crack, we stopped and examined it again. It appeared that the lake bed surface had simply sunk down in a long jagged line that extended for several miles. This was an abnormal feature to encounter on the lake bed. Normally the surface was perfectly flat, with only minor sinkholes or surface cracks. We routinely landed aircraft on all but a few isolated areas of the lake bed during the late fifties and early sixties. Now, all of a sudden, we were rudely faced with a hazard that could tear the landing gear off an aircraft if a pilot should land without noticing the crack before touchdown. As the years progressed, we began noticing more large cracks and larger sinkholes. In the late sixties and early seventies, it became necessary to fill and smooth the cracks to ensure that the lake bed was safe for use by aircraft. By the eighties, the cracks and potholes were so numerous that only the marked runways on the lake bed were repaired. It became impossible to fill all the cracks and holes in that huge lake bed.

During this time period, we began to suspect that the entire lake bed and surrounding area were slowly sinking because of the rapid depletion of the normal groundwater. The areas to the south and west of Edwards were being heavily farmed by alfalfa ranchers, who were harvesting three and four crops per year. This was accomplished using groundwater for irrigation, which was pumped at a tremendous rate. Each year, the ranchers would have to dig deeper and deeper for an adequate supply of water. From the mid fifties until the late seventies, the water table sank as much as 600 feet. During the late seventies, the alfalfa ranchers began going out of business. They couldn't afford to pump the volume of water they needed from that depth. The Air Force Flight Test Center has recently taken steps to minimize lake bed damage by restricting the pumping and use of local groundwater and relying instead on the state water system. Without that action, the lake bed would eventually have been irreparably damaged.

The winter rains that used to flood the lake bed and wash away any surface irregularities, man-made or otherwise, leaving the surface glassy smooth, now severely damage the surface. Instead of remaining on the surface and washing back and forth across the lake with the shifting winds, the water soaks rapidly into the lake bed, creating large sinkholes. It will be many, many years before the groundwater is replenished. Until then, the lake bed will continue to sink and crack, destroying a beautiful natural landing area. I'm not a raving environmentalist, but the destruction of the Edwards lake bed has occurred in my lifetime, a short time compared with the aeons that the lake bed remained in pristine condition before man arrived.

THE FIRST M2-F1 FREE FLIGHT

A test pilot likes everything to go smoothly on the day of a first flight. Things were not going too well on the morning of August 16, 1963. We were preparing for the first M2-F1 air-towed flight to altitude and the subsequent free flight after towline release. Our plan was to take off on Runway 35 on the south lake bed and fly north until reaching the north shoreline. At that point, we would begin a shallow turn to the west and continue in a wide circle around the northwest portion of the lake bed, while climbing to our desired release altitude of 7,500 feet above sea level. This would be roughly 5,200 feet above the lake bed.

We would maintain that altitude until we crossed the north end of Runway 18-36 on the north lake bed. At that point I would release and turn to line up with that runway, landing to the south. From that release altitude, I should land about 2.5 miles down the runway, assuming our predicted lift-to-drag ratio of 3 was reasonably accurate. The flight plan was relatively simple, and we hoped we had anticipated any probable contingency. We were staying close to the lake bed throughout the climb pattern in case the towline broke, and the plan did not require any significant maneuvering in free flight in case the M2-F1's flying qualities were unsatisfactory.

The plan was to take off shortly after daybreak to take advantage of the light winds and low turbulence typical of early summer mornings in the desert. All of the participants were scheduled to come to work at 4 o'clock in the morning. Our plan involved towing the M2-F1 down to the south lake bed takeoff position with the Pontiac. The C-47 would fly down to the south lake bed and park in position for the takeoff. We would start our preflight checks at that location and then hook up the towline.

Our first deviation from the plan occurred when the C-47 pilot failed to show up on time. We waited and waited, but he didn't show up. Don Mallick was the first pilot to show up while we were waiting, so we pressed him into service as the C-47 copilot. The C-47 finally arrived at the south lake bed about 45 minutes late, but the weather was still acceptable for flight. The crew and I had completed all of our preflight checks, so we were ready to hook up the towline and go. The crew hooked up the towline to the M2-F1 and then pulled on the line while I released the tow hook to ensure that my release was working properly. The crew did the same on the C-47. Vic Horton was the observer in the C-47 and the individual who would release the towline from the C-47 in case of some unanticipated problem.

When those checks were completed, the crew installed and locked my canopy and the C-47 crew started their engines. We were ready to go as soon as they completed their engine checks. The C-47 crew called the Edwards Tower for takeoff clearance and requested that they close the base during our takeoff and initial climb to eliminate any conflicting traffic in case I had to release early or we had some other problem requiring a change in our flight plan.

Once we received takeoff clearance, Mallick checked with me to see if I was ready to go. After I confirmed that I was, we started rolling.

As we accelerated through 50 knots, I pulled the nose up in preparation for a liftoff at 70 knots. Just as I prepared to rotate for liftoff, the towline released from the C-47. I let the nose back down and rolled to a stop. I wasn't sure what had happened, whether the towline had broken or whether it had been inadvertently released. The C-47 pilot aborted the takeoff and taxied back to our starting position. The ground crew hooked me up to the Pontiac and towed me back to the starting position. A quick examination of the towline revealed that it was intact. It soon became obvious that the towline had somehow pulled out of the C-47 hook. A close examination of the hook confirmed that it was intact and operating properly. The crew then attempted to reinstall the towline metal end ring in the tow hook and found that the ring was too thick to allow the hook to close and lock properly. We now had encountered a second deviation from our plan.

I was beginning to get a little concerned. Vic Horton suggested that he take the metal ring back to the machine shop and grind it down. This would take at least half an hour to accomplish, since it was over 10 miles back to NASA. I was concerned about the wind picking up and the air becoming turbulent with the rapid increase in temperature as the morning progressed. During a three-hour period on a summer morning, the temperature could rise more than 40 degrees from an early morning low of 60 degrees F to over 100 degrees F by 9 o'clock. The atmosphere quickly becomes dynamic under those conditions.

I reluctantly agreed to continue trying for a flight. While I was waiting for Horton to get back, I thought about the premature towline release on the previous takeoff attempt. I realized that I was lucky that the rope released before liftoff. We had a predicted dead-man zone from roughly 100 to 200 feet after takeoff. If the towline broke while climbing through that altitude band, I would not be able to maintain enough airspeed to make a landing flare. The M2-F1 decelerated rapidly when the towline released, approximately 7 knots per second. The only way to maintain airspeed was to dump the nose over and try to establish a stabilized glide at roughly a negative 25-degree glide slope. It was impossible to accomplish all of this below a 200-foot altitude. The aircraft would not have sufficient energy (airspeed) to make a successful flare. It would hit the ground hard, too hard.

Defining the dead-man zone was part of the preparation for the first flight. Anticipation of potential problems is a major part of the preparation for any test flight, but it is even more important for a first flight. You must examine everything and anything involved in the flight and anticipate problems and the effects of those problems. In flight test or flight research, everyone involved is looking for things that can go wrong. There should be no surprises during flight, because someone should have identified any potential problem and either taken steps to eliminate the problem or defined procedures to cope with the problem if it does occur.

Simulation is one of the many tools used to search out potential problems. We identified and defined the dead-man zone using simulation. The test pilot has to familiarize himself with all of the identified potential problems and search for more, since these problems can have a deadly effect on him. The pilot must be prepared to cope with any possible problem on his own. He must have an escape plan for any conceivable emergency if he hopes to become an old test pilot.

Horton and the M2-F1 crew chief came roaring back across the lake bed about half an hour later. They checked the ring fit on the C-47 tow hook. It fit correctly. We were ready to try again. It required another half an hour to conduct all the preflight checks, start the C-47 engines, and get the towline stretched out. Finally, we were ready to go. Don Mallick, the C-47 copilot, called the tower for takeoff clearance and advised the tower to close the field again until we got airborne and clear of the main runway. The tower advised us that the surface winds were light and variable and then cleared us for takeoff.

As we started rolling, Horton called and requested that I confirm that I had the landing rocket arm switch on and had a green light, and that my inverter was on. I confirmed each of those items. He then called for "data on." I lifted off as we accelerated through 75 knots and immediately climbed to a position above the C-47. I had previously briefed the C-47 pilot to stay on the ground until I was in position above the C-47. This was to ensure that the strength of the tow plane wake was minimized as I climbed through it. The strength of the wake is proportional to the amount of lift being generated by the aircraft. If the C-47 is still on the ground, it is not generating a significant amount of lift, and the aircraft wake is not noticeably turbulent.

When I reached a comfortable towing position above the C-47, I called and cleared them to take off. They continued to accelerate after lifting off until we reached 110 knots and then began to climb. I wanted as much speed as possible at low altitude in case the rope broke, to enable me to flare the M2-F1 for a landing. The M2-F1 decelerated rapidly in level flight, and I did not want to stall out 50 or 100 feet in the air. When my airspeed stabilized, I called Mallick for an airspeed check. He confirmed they were indicating 110 knots. As we continued to climb, I varied my position behind the C-47 to achieve the most comfortable position in terms of control forces and visibility. I definitely wanted to stay above the C-47, since the wake of an aircraft tends to trail below the aircraft and descend slightly with time.

The only way to maintain visual contact with the tow plane while flying above it was to use the nose window. This was not the most desirable window, since I had to fly with my head down to see through that window, but it was adequate. The control forces were light, so I had no problem maintaining that position above the tow plane. As we reached the north shoreline, I called the tower and cleared them to open the field. Things were going quite well, and I didn't anticipate any emergency requiring a premature release and landing. The C-47 pilot began a gentle left-hand turn to parallel the shoreline as we had planned and continued climbing.

Vic Horton noticed an aircraft off to our left and asked to confirm the sighting. I saw it as Horton called out its position. It appeared to be an F-104 climbing toward us. I suggested that the tower be called to get that aircraft out of our area. I was afraid that he might be trying to get a close look at the M2-F1 and might not notice the towline between me and the C-47 and accidentally collide with it. That could be catastrophic. The snatch load on the towline before it broke could tear the fragile hull of the M2-F1 apart.

As we continued to climb, we would occasionally encounter some light turbulence. The M2-F1 reacted rapidly and quite noticeably to turbulence, primarily because of the high dihedral effect and the lack of roll damping. The small amount of turbulence-induced sideslip would cause the M2-F1 to roll off abruptly. It was not a significant problem, but it was annoying. I suggested we slow down a bit since the M2-F1's response to turbulence decreased as the airspeed decreased. So far, the mission was proceeding quite well. I called for a check of the surface winds and was informed that they were light, from the south at 4 to 5 knots. The weather was cooperating nicely.

I noticed some buffet as we continued the climb. I could feel it in the stick as a small vibration. I checked the outboard elevons for any vibration in my mirror, but they were not moving at all. I decided the buffet must be from vibration of the rear upper flap. We had seen those flaps vibrating during the wind tunnel tests. We added some small turning vanes at the trailing edge of the fuselage to direct air into the blunt base area in an attempt to alleviate the buffeting. The vanes helped but did not eliminate the buffeting completely. That residual buffet was undoubtedly what I was sensing. Horton asked if Fred Haise, flying in the T-37 chase plane, should move in to check the flaps. I said yes, but I warned Haise not to get directly behind the M2-F1 in case the towline should break. He would get a faceful of the M2-F1 if that happened. Haise verified that the rear upper flaps were oscillating at a small amplitude.

As we continued on around the turn and began heading back toward the north end of the lake, I noticed that an aircraft was landing on Runway 23. That runway crossed the north lake bed from northeast to southwest. It was inside our approved operating area for this flight. I suggested Bruce Peterson call the tower and have them close that runway. Peterson was the ground controller for the flight. I didn't want any conflicting traffic on the north lake bed, since I couldn't predict where the M2-F1 might go after release from the towline. We were pretty confident that it would go where I wanted it to go, but there was always the possibility of a control problem on a highly unconventional flight vehicle such as the lifting body. As we approached the north shore, I called all stations to stand by for towline release on this pass. The radio transcript includes the following conversation:

MALLICK (C-47 copilot): Okay, Milt, do you want to set up on this one [pass] for the drop?

THOMPSON: Affirm.

MALLICK: Okay, swing a little bit wide coming around here.

THOMPSON: Rog.

HORTON (C-47 observer): Milt, I'll give you a two-minute and one-minute count. Will you give a thirty-second, twenty-second, and ten-second countdown to release?

THOMPSON: Okay.

MALLICK: Milt, at one minute, do you still want us to go to one-hundred-five knots for drop?

THOMPSON: Yes, go back up to one hundred five just before I release.

MALLICK: Okay.

NASA 6 (ground control): Zero one seven [call number of C-47], this is NASA Six control. Joe Vensel [chief of flight operations] says the winds are from the south at zero one . . .

THOMPSON: I didn't read the last on that.

NASA 6: I think it's four knots. Is that correct NASA One?

MALLICK: I'll get it from the tower, Milt.

NASA 6: Winds appear to be very light down here. Very light.

THOMPSON: Rog.

NASA 6: Zero one seven, the winds are from the south at five to eight knots on the lake bed.

THOMPSON: Rog, we got that.

HORTON: Two minutes until release.

THOMPSON: Don, when I call thirty seconds to release, you can start accelerating a left turn at one-hundred-five knots.

MALLICK: Okay, Milt. The winds are one-ninety at seven.

THOMPSON: Rog. It's going to work out just about right, altitude-wise.

HORTON: One minute until release. Turn your inverter on. Milt, arm switch on [landing rocket arm switch].

THOMPSON: Inverter on. Oscillograph switch is on. Arm switch is on.

HORTON: You can take the count from here, Milt. Give him a thirty-second and turn your data on now.

THOMPSON: Okay, thirty seconds. Data coming on. Start accelerating.

MALLICK: Rog, accelerating. Throttling back.

THOMPSON: Okay. Camera is coming on. Remind me to turn this off. Arm is on. Release. Okay, camera is off. Starting to turn now. [Turning the M2-F1 toward the lake bed to land on Runway 18.]

HORTON: Turn your camera on, Milt.

THOMPSON: I didn't have it on for that but the field looks pretty good. Camera is coming on now. Okay, darn, left it on for a little bit. I'm going to leave it off until just about the time I start to flare. [We had very limited film in the camera.] Going off. Data on. Little rough in here.

NASA 6: Okay, Milt, fifty feet, twenty feet, ten feet, five feet, you're on. Real nice. Good show, Milt.

The flight lasted under 2 minutes. My average rate of descent was roughly 4,000 feet per minute. The flight went smoothly. I turned in toward the lake bed after release and then lined up on Runway 18. I had preselected the release altitude and location so that I would land about halfway down the runway if I flew straight ahead after I turned in to line up on the runway. That would ensure that I didn't have to maneuver to achieve a safe landing location. There was no guarantee that I would be able to maneuver after release, since no one had ever flown a lifting body.

The purpose of this flight was to determine whether we could maneuver and land the vehicle. The wind tunnel data predicted we could. This flight would confirm or deny those predictions. I began maneuvering mildly to determine how the vehicle behaved. It behaved well. I could turn easily and roll out on a desired heading quite accurately. Next, I tried varying airspeed by changing pitch attitude. Again, the vehicle responded well. It was now time to set up for the landing flare. I pushed over to pick up 120 knots and rolled the vehicle gingerly to assess its handling qualities at that speed. Everything felt good. I began pulling the nose up to verify I had good control response for the final flare. Again, everything felt good.

At 400 feet above the lake bed I began the final flare. The vehicle responded well, so well, in fact, that I delayed completing the flare until I got closer to the lake bed. At 200 feet above the lake bed, I resumed the flare maneuver and came level about 10 feet above the lake bed. From that height, I allowed the vehicle to settle on down to touchdown. It was a soft touchdown at approximately 75 knots. I let the nose down and rolled to a stop with only minimal braking. The flight was a success. The vehicle flew nicely, and there were no surprises. That's the way you would like every test flight to conclude. Unfortunately, some don't.

Everyone at Dryden was outside watching that flight. There were employees on the roof of the building, on the aircraft parking ramp, on the taxiway, and out on the edge of the lake bed. A good share of the Flight Test Center personnel and the contractor tenants were also out on the ramp and taxiways watching the flight. They had watched our ground-towing operations for months, as we roared up and down the lake bed creating huge clouds of dust. After all that, they weren't about to miss the first flight. Most of the people working at Edwards were somewhat blasé about watching new aircraft fly, since they had seen so many unusual aircraft pass overhead, but the M2-F1 was something else—a flying machine without wings, a radical shape symbolizing a new generation of flying machines that might someday fly into and back from space.

It was the usual practice that family members did not watch test flights, but for the first M2-F1 air tow my wife, Therese; daughter, Brett; sons Eric, Milt Junior, and Peter; two of my sisters; and my mother, Alma, were on hand. They were standing on the edge of the lake bed. Two engineers were standing behind them as the M2-F1 descended. Eric remembered that one said, "It's gonna come out of the sky like a brick." The other engineer responded, "No, it will flare just before it hits."

I spent the week after the first air-towed flight supporting the X-15 program. I flew a weather flight in an F-104 on one day and then flew up to Beatty, Nevada, to monitor and control the first segment of X-15 flight 3-22-35. During the early portion of the X-15 program, we were unable to transfer data from one tracking station to another, so we required a control-room operation at Beatty as well as at Edwards. That added a lot of complexity to the operation.

It wasn't until August 27, 1963, that we made the next M2-F1 flights, a pair of ground tows. The following day we made the second air-tow flight. The third followed on August 29 and two more the next day. These flights were devoted to data gathering. We obtained data on the stability, control, and handling qualities of the vehicle in different flight conditions. We also gathered data on its performance characteristics.

Two research flights made on September 3 marked the introduction of the M2-F1 to the press. *Aviation Week and Space Technology* and *Missiles and Rockets* both carried articles on the program. One reporter likened the vehicle to a flying bathtub. Although a number of people at NASA headquarters were aware of our homebuilt project, they did not include NASA administrator James E. Webb. His introduction to the M2-F1 came when a congressman who had read a newspaper story about the flights asked him about the program. A few feathers were ruffled, but the low cost of the program smoothed things over. A few days later, Dr. Clarence A. Syvertson of Ames Research Center said at a seminar held at Edwards Air Force Base that lifting bodies could satisfy requirements for maneuvering reentry from Earth orbit and for a manned mission to Mars.

The positive publicity NASA gained from the M2-F1 attracted the attention of some people in the Dyna-Soar program. It was obvious that Dyna-Soar was in deep trouble politically and financially. Program management was searching for ways to keep the program alive. One proposal was to build a low-cost Dyna-Soar vehicle similar to NASA's M2-F1 wooden lifting body. Proponents of this option were impressed by the news-media attention given to the M2-F1. NASA was getting a lot of visibility for a small investment. The idea of a low-cost Dyna-Soar was squelched as soon as various subsystem engineers began advocating the inclusion of their subsystems into the low-cost vehicle to obtain flight experience. It quickly became obvious that neither the Air Force nor Boeing program managers could build a real low-cost Dyna-Soar vehicle. Clearly, they did not have the same mentality that drove the Skunk Works group at Lockheed to put together innovative aircraft on tight schedules and under budget. We discovered this when the major aerospace contractor we contacted gave us an estimate of $150,000 to build the M2-F1. We did it for a fifth of that price. I have subsequently observed attempts by other aerospace organizations to produce low-cost vehicles. The majority of the attempts were failures. It is not easy to develop an organization that can create truly low-cost vehicles. Lockheed's Skunk Works will not be seriously challenged for many years.

The September 3 flights brought a temporary close to M2-F1 flights. I spent the first week of September on travel to Worcester, Massachusetts, and Grand Rapids, Michigan. At Worcester, I was being measured for an X-15 pressure suit. At Grand Rapids, I was evaluating a new X-15 cockpit instrument display

Milton O. Thompson in the Paresev 1-A on the ramp at Dryden, with the Piper Super Cub tow plane in the background. The Paresev had been rebuilt following Bruce Peterson's crash. Courtesy of National Aeronautics and Space Administration.

Left: Milt Thompson preparing for a research flight in the Paresev 1-A. The Paresev was built at an original cost of $4,280; the framework was laid out with chalk on the welding shop floor. Courtesy of National Aeronautics and Space Administration.

Right: The Paresev 1-A in flight above Rogers Dry Lake. Originally designed to test the concept of the Rogallo wing—a steerable, wing-shaped parachute—the Paresev had control problems that prevented the use of a similar design to land the Gemini spacecraft. Courtesy of National Aeronautics and Space Administration.

The M2-F1 (left) with the heavyweight M2-F2 on the ramp at Dryden. The wooden M2-F1 had fixed landing gear and horizontal control surfaces attached to the fins. The all-metal M2-F2 had retractable landing gear and a configuration closer to that of an orbital lifting body. Courtesy of National Aeronautics and Space Administration.

The M2-F2 under construction at Northrop in Hawthorne. The design called for extra weight in the nose to correct the center of balance, but rather than just adding lead ballast, the builders reinforced the cockpit structure. Courtesy of National Aeronautics and Space Administration.

The M2-F2's first free flight, July 12, 1966. Thompson had to let go of the controls a few seconds before landing to stop a pilot-induced oscillation. Courtesy of National Aeronautics and Space Administration.

The HL-10 with the X-15A-2 at Dryden. The X-15A-2, tested at the same time as the first lifting bodies, could reach speeds of Mach 7. The HL-10, like the other lifting bodies, was a technology demonstrator designed to show the practicality of the lifting-body concept at speeds up to Mach 2. Courtesy of National Aeronautics and Space Administration.

The HL-10 following modification of the tip fins. The handling problems on its first glide flight were traced to flow separation over the fins. Courtesy of National Aeronautics and Space Administration.

Bruce Peterson after piloting the first flight in the HL-10. The vehicle proved to be nearly uncontrollable, requiring extensive wind tunnel tests to develop a solution. Courtesy of National Aeronautics and Space Administration.

Above: The M2-F2 lifting body following the crash of May 10, 1967. The vehicle came to a stop upside down, resting on the left fin and pilot's seat. Pilot Bruce Peterson was pulled from the cockpit, on the right. Courtesy of National Aeronautics and Space Administration.

Below: The resurrected M-2. After its crash, the M2-F2 was rebuilt as the M2-F3. The control problems that had plagued the vehicle from its first flight were cured by the addition of a center fin seen here with pilot Bill Dana's name substituted for the standard "NASA" label. Photograph courtesy of Jack Kolf.

The M2-F3 during preflight ground tests. Moment-of-inertia hinges are attached to the vehicle's side. Given the earlier problems with stability, training wheels seemed appropriate. Courtesy of National Aeronautics and Space Administration and Jack Kolf.

A flock of lifting bodies. From left to right: the X-24A, M2-F3, and HL-10. Courtesy of National Aeronautics and Space Administration.

Test pilots at play. The nefarious repainting of vehicles was a popular pastime at Dryden. This photograph of Chuck Archie (left) and Jerry Gentry was retouched to add Gentry's Captain Midnight costume and the Air Force markings on the previously NASA-only HL-10. Photograph courtesy of Jack Kolf.

Above: The X-24A, the third of the heavyweight lifting bodies. Designed through the Air Force, it was the ugliest of the first generation of lifting bodies and was likened to a finned potato. Courtesy of National Aeronautics and Space Administration.

Below: Maj. Mike Love and the X-24B. The original X-24A fuselage was nearly buried within a new outer skin, including the extended nose, stub wings, and flat underside. Only the canopy, three fins, and rear fuselage remained visible. Courtesy of National Aeronautics and Space Administration.

Above: The M2-F1 lifting body and the space shuttle *Enterprise* at Dryden. Courtesy of National Aeronautics and Space Administration.

Below: The X-38 lifting body, a test vehicle for the space station CRV (crew return vehicle). To serve as a lifeboat should the space station or its crew suffer an emergency, the CRV is designed to reenter the atmosphere automatically and make a precise landing using a steerable parasail. Courtesy of National Aeronautics and Space Administration.

An artist's conception of the X-33 after launch from Edwards Air Force Base. This lifting body is designed as a technology demonstrator for a single-stage-to-orbit vehicle. Courtesy of Lockheed Martin and the Edwards Air Force Base History Office.

The full-scale Venture Star in orbit. The vehicle is designed as a space shuttle replacement, with a lower launch cost than any existing booster. Courtesy of Lockheed Martin and the Edwards Air Force Base History Office.

panel. My priorities had changed. Instead of devoting most of my time to the M2-F1, I would now have to prepare for my first X-15 flight, which tentatively would take place in the latter part of October. I had to attend X-15 ground school, fly the X-15 simulator, and practice X-15 unpowered approaches in one of our F-104 aircraft. I flew F-104s almost every day, practicing X-15 approaches during the remainder of September. I also flew X-15 chase missions and served as the mission controller. It was a rather hectic month.

It wasn't until October 7 that I made my next M2-F1 flight. We were able to fly the M2-F1 only by lengthening our workday. I would come to work about 4:30 in the morning and begin flying the M2-F1 at daybreak. I would then devote the remainder of the day to X-15 activities. I was able to make one flight each on October 7, 9, 15, and 23, and two on October 25. On October 29, 1963, I made my first X-15 flight. What a tremendous change: from a wooden glider being towed by a C-47 to a steel airplane being dropped from a B-52 bomber, from 100 knots to 4,000 knots, from 12,000 feet maximum altitude to 200,000 and 300,000 feet altitude, from glider to rocket ship.

In November, I continued to devote most of my time to X-15 activities but was able to make three air-towed flights on November 8. The lifting-body advocates were chomping at the bit to continue flying the M2-F1. Between November 12 and 14, and again on December 2, I began a series of ground-tow flights to check out some other pilots in the M2-F1: Bruce Peterson, Don Mallick, and Chuck Yeager. Peterson was scheduled to take over eventually as the M2-F1 project pilot, and Mallick would be his backup. Chuck Yeager wanted to fly the M2-F1 to evaluate the potential use of a lifting-body trainer for the Aerospace Research Pilots School that he was commanding at that time. Each of the three pilots made eight to fourteen ground-tow flights per day. The one problem was the huge dust clouds raised by the Pontiac as it raced back and forth across the lake bed. It was hard to judge your height while flying in the dust, much harder than with fog, as the dust got thicker closer to the ground, not thinner as with fog.

I was scheduled to move out of the M2-F1 program so that I could devote all my time to the X-15. It didn't quite work out that way though. I continued to fly both the X-15 and the M2-F1 for some time. I made another X-15 flight on November 27.

On December 3, Yeager and Peterson made their first air-towed flights. On Peterson's second flight, the wheels broke off the M2-F1 on landing. Peterson contended that the wheels broke off because of the high viscosity of the oil in the landing-gear shock struts. The temperature that morning was below freezing on the ground. Obviously the temperature's effect on the oil in the shocks was a contributing factor, but the onboard, over-the-shoulder camera revealed another cause: an incomplete flare maneuver. Peterson failed to arrest the vehicle's vertical velocity before touchdown. The over-the-shoulder camera sequence made me cringe as I watched the lake bed surface rush up toward the pilot. Prior to flare, the M2-F1 was descending at about 5,000 feet per minute, or about 90 feet per second. The ground was coming up fast in the last few seconds. In the film, it is obvious that the rate of descent was still very high when the M2-F1 touched down. The vehicle slowed down rapidly after touchdown as it was sliding out on the landing-gear struts. In the over-the-shoulder film, as the vehicle slides to a stop, the wheels can be seen bouncing along the lake bed ahead of the M2-F1. It really was a hilarious scene. Those wheels seemed to roll on for miles.

The M2-F1 went into the hangar for repairs, while I spent the next week in Dayton, Ohio, at Wright-Patterson Air Force Base undergoing some Dyna-Soar pressure-suit tests. While I was in Ohio, Chuck Yeager ejected from the NF-104A, and the Dyna-Soar program was canceled. With the Dyna-Soar gone, it was clear that it would be a long time before a manned spacecraft made a lifting reentry and horizontal landing. The M2-F1 and the emerging heavy-weight lifting-body programs would have to keep that dream alive, building experience and data until the political support developed to begin such a program.

OWEN BILLETER

The M2-F1 was repaired by late January 1964, and Peterson and I began a series of air-towed flights. We also finished checking out Yeager (who had now recovered from the burns suffered in the NF-104A ejection) and Mallick. We made ground-tow checkouts of Donald Sorlie and Jim Wood in February. Between April and August 1964, Peterson and I made relatively few air-towed flights in the M2-F1. Operations picked up again in early 1965.

We occasionally were frustrated on our early morning flight attempts. Our crew chief, Owen Billeter, wouldn't show up on time or would fail to show up altogether. Billeter was single and a real carouser. He lived in Rosamond, and Rosamond at that time was a pretty tough town. It had only four bars, but they were busy around the clock. They didn't serve booze between two and six in the morning, but the poker games ran nonstop. The bars were normally frequented by ranchers, prospectors, and gamblers. Fights were a part of the normal nightly activity. There was no local policeman or sheriff, so the bartenders or bar owners would have to confront the drunk and disorderly patron. Quite often, this resulted in a fight.

Billeter spent most of his time away from work in these bars. He also tended to become drunk and disorderly. He often bore the scars of battle: a black eye, cuts and bruises, a gap in his teeth, and a mashed and swollen lip. He would occasionally show up for an early morning flight with one or more such scars and an obvious hangover. It hurt just to look at him. If he made it to work, he would work diligently and cheerfully, but you could sense the pain. He seemed to thrive on this lifestyle. It was often humorous to watch the look on a visitor's face when he saw Billeter report to work at 4:30 in the morning. We were a little embarrassed, since we were trying to create an impression of a highly competent and professional organization. Somehow, one doesn't look very professional with a black eye and split lip.

Billeter finally left NASA in the late sixties and went to Vietnam as a civilian to work for the Air Force maintaining F-4 fighter-bombers. I would hear from him every couple of years for the next ten years, but then the phone calls and letters stopped. I often wonder whatever happened to him. He was a conscientious crew chief, except on those infrequent occasions when his nightlife overwhelmed his daylife. He took great care of me in the M2-F1.

DANA'S CHECKOUT IN A THUNDERSTORM AND GENTRY'S SLOW ROLL

In preparation for the heavyweight lifting-body program, we used the M2-F1 to check out several additional pilots. Bill Dana, Jerry Gentry, and Donald Sorlie made ground tows and later air-towed flights in the M2-F1. Dana's one and only air-tow flight in the M2-F1 was made on July 16, 1965. It wasn't the best of days for a flight; we had arrived in the early morning but had to wait until a thunderstorm cleared the area. The cloud cover was at 10,000 feet, high enough for the C-47 tow plane to fly, but it was raining, and when the lake bed became wet it turned into a heavy, thick mud. We were walking around with half an inch of mud on our boots.

Dana wanted to cancel the checkout, but we went ahead anyway. I made a flight to check out the vehicle, then we put Dana in the cockpit for the checkout. When we were done, the C-47 started its engines and towed him aloft. The M2-F1 lacked a trim adjustment, so during the long climb to release altitude, Dana had to apply forward pressure. It was not a large amount of force, but it was tiring. Dana released the towline, then pitched down into the 30-degree glide angle. What he did not know was that about a quart of rainwater had collected in the vehicle. As he pitched down, the water flowed forward and pooled in the lower window. The lake bed looked 10 miles away, and it was hard for Dana to judge his height. He knew he had only a few seconds to make the flare maneuver. The M2-F1 was airworthy but not watertight, and the water quickly drained away. Dana made a successful landing.

Now it was Gentry's turn. He climbed into the cockpit, then we lowered the canopy and hooked up the towline. Gentry radioed that he was ready, and the C-47 started its takeoff roll. At 75 knots Gentry lifted the M2-F1 off the lake bed and got into position slightly above the tow plane. As we watched, the M2-F1 slowly began to oscillate through a bank angle of 20 to 30 degrees at a frequency of about 1 cycle per second. The motions seemed to stop once or twice, then increased rapidly to an amplitude greater than 180 degrees. I radioed Gentry to "Level your wings, level your wings." As the M2-F1 continued to roll until it was nearly inverted, I started to panic. I radioed, "Release! Release!" and then, as the M2-F1 went inverted, I yelled, "Eject! Eject!" At that moment both Gentry and the C-47 observer released the towline. Gentry was upside down with a nose-high attitude at 400 feet and an airspeed of approximately 100 knots. The probability of recovery from that condition was zero. For a normal landing, the flare was initiated at about 120 knots. It was theoretically impossible to get the nose down to pick up speed. Gentry ignored theory and simply completed the barrel roll level as he flared. The M2-F1 landed hard, but both it and Gentry seemed okay. Dana and I were both in shock.

Gentry thought he knew what had happened and wanted to give it another try. We were still in shock and agreed to make another attempt. We towed the M2-F1 back to the starting point and then noticed that it was listing slightly. As the crew chief began to inspect the landing gear, a radio call came in from Joe Walker: "Get that thing back in the hangar, now!" Walker was the only person with any common sense that day. We later found that the impact had caused severe damage to the hull structure at the landing-gear attachment points. Had we made a second attempt, the gear could have fallen off. While the M2-F1 was under repair, we decided to halt the pilot checkout.

Following Gentry's slow roll in the M2-F1, we decided to check him out in a sailplane and give him some towing experience before making another attempt. Gentry and I went to Tehachapi in the Aero Commander, accompanied by Paul Bikle and a visiting Air Force general. Bikle thought that I could demonstrate the problem of pilot-induced oscillations (PIO) on tow in a sailplane to prevent Gentry from having another PIO in the M2-F1. We rented a Schweizer 232 for the demonstration. Fred Harris was the tow pilot. I put Gentry in the front cockpit, and I climbed in the back. We were lined up on the runway to take off headed east. The wind was from the east at about 10 miles per hour.

I told Gentry to fly the sailplane and I would tell him what to do. As we accelerated down the runway, the wind shifted abruptly and increased in speed. The sailplane was being blown off the runway before we got airborne. The runway was narrow, about 50 feet wide, and the shoulders were soft. I decided to take command of the sailplane, since Gentry had not flown one previously. I told him, "I've got it," and attempted to take over the controls. Gentry wasn't about to admit defeat. He wouldn't let go of the controls. He said, "No, I've got it," and continued to apply control inputs. By this time we were at the edge of the runway.

I said, "Let go, Jerry," and hauled back on the stick. We lifted off abruptly and drifted off the edge of the runway. The wind had shifted almost 120 degrees and was now a quartering tailwind. This type of wind shift was not uncommon—opposite directions at each end of the runway. The cause was an unusual circulation pattern in that mountain pass. Gentry was still fighting me for the controls after liftoff, but he finally gave up. Once we were stabilized on tow, I gave the controls back to him for his PIO training. The incident heightened my concern about his ability to handle a PIO situation, but I was willing to give him another chance.

While I was checking Gentry out, Bikle and the general decided to make a sailplane flight just for the pleasure of it. They each rented their own sailplane. After Gentry and I landed from our dual flight, I also rented another sailplane and went back up to enjoy the thermal activity. Gentry came up on a solo flight to obtain some more flight time. It was a weekday, and we had the airport to ourselves. We were soon making simulated gunnery runs on each other as we climbed to higher altitudes.

MY ENCOUNTER WITH A DISNEY PRODUCER

It was during this trip to Tehachapi that I was introduced to Ken Nelson, an independent movie producer who was making a television film for Walt Disney. It was called *The Boy Who Soared with Condors* and starred Christopher Jury and Margaret Birsner, who were actually young glider pilots.

The story opens with Jury watching California condors soaring on air currents. As he watches, a sailplane lands nearby and Birsner climbs out. She asks him to verify the location of the landing so she can earn her Silver C distance badge, which requires a flight of 31 miles. Jury enrolls in the Tehachapi Soaring School, where he learns to fly sailplanes, then successfully makes a series of ever longer cross-country flights. He finally tries for soaring's highest award, the Diamond C, which requires an altitude of 16,000 feet and a distance of 311 miles.

The flying scenes were being done out of Tehachapi, and Jury's tow pilot and instructor in the film was played by Fred Harris, who had done the tow work for us on the Paresev project.

I soon found myself and the M2-F1 being written into the script. In the film, Harris introduces me to Jury and Birsner and I invite them to visit Dryden and watch an M2-F1 flight. The Disney crew filmed scenes of me showing them the X-15 and M2-F1, and of Birsner sitting in an F-104. After arrangements were made, I made an air-towed flight on August 31, 1965, for the Disney film. The C-47 tow aircraft was flown by Bruce Peterson and Fred Haise. The film crew shot footage of me climbing into the M2-F1, wiggling the elevons, being towed aloft, and then gliding down to a landing while Jury and Birsner watched.

The film was shown on *Walt Disney's Wonderful World of Color* on February 19, 1967, with a rerun on May 28, 1967. This was the first time the public had a closeup look at the lifting-body program.

THE END OF THE M2-F1 PROGRAM

Following Jerry Gentry's PIO in the M2-F1, we halted the pilot checkouts for nearly a year. It wasn't until April 22, 1966, that Bruce Peterson, Fred Haise, and Joe Engle made ground-tow flights. Three weeks before, both Haise, a NASA research pilot, and Engle, an Air Force X-15 pilot, had been selected for astronaut training. Flying the M2-F1 would provide experience for later flights in a lifting spacecraft. Haise flew aboard Apollo 13, and both he and Engle made approach and landing tests of the space shuttle *Enterprise* at Edwards in 1977. Engle flew into space aboard the shuttle on STS-2 in November 1981 and STS-51I in August 1985.

Now it was again Gentry's turn. On July 21 and August 10 he made successful ground-tow flights. On August 16, 1966, exactly a year and a month after his slow roll in the M2-F1, Gentry was to make his second air-towed flight. On hand were Bill Dana, I, and the same cast of characters from that first attempt. We did the preflight preparations, Gentry radioed he was ready, and the C-47 began its takeoff roll. Gentry made a perfect takeoff and moved into position above the C-47. He stabilized in the correct position, and I started breathing again.

Then Dana and I both noticed the M2-F1 was swinging from side to side ever so slightly. As we watched, the amplitude rapidly built up to greater than 180 degrees and the M2-F1 was inverted. It was an exact replay of Gentry's first PIO. I was so shocked that I could only yell, "Eject! Eject!" The previous experience must have done Gentry some good, as he just released the towline, fired the landing rocket, and flared for a landing. The touchdown was smooth, and he did not damage the M2-F1. In fact, it was a beautiful landing. Dana and I, on the other hand, couldn't even talk. Never in my life have I been so scared as during those two incidents.

Gentry was now in deep trouble with the general commanding Edwards. The general called Paul Bikle and informed him that he was going to recommend that Gentry be taken off the lifting-body program. The general was extremely conservative regarding flight safety. Many believed he would rather have stopped all flying than have an accident. In the early days at Edwards, flight testing had been pretty loose, and the accident record showed it. During the fifties, things were tightened up considerably, and in the early sixties there was a period when Flight Test Center generals were getting fired for having an accident. That was the cause for the general's ultraconservatism regarding flight safety.

My attitude was that a flight test center must fly to determine the potential problems or sources of accidents in a new airplane and then correct them. You are not going to accomplish your primary objective as a test center commander if you are overly conservative. That did not mean you should take unreasonable risks or use poor judgment. It simply meant that you should do a thorough job of simulation, analysis, and training, and then fly even though there may be some risk. You cannot eliminate all risk. Someone has to bite the bullet and fly.

Bikle called me down to his office and told me that we had to talk to the general to keep Gentry on the program. Bikle's defense of Gentry was something of a surprise because he had just chewed Gentry out the day before for not recognizing the problem and taking the corrective action. But that was Bikle's nature—to chastise and then defend.

Bikle called the general and scheduled a meeting for the following day. We drove over to the headquarters building the next day and went up to his office. The general had Col. Donald Sorlie, chief of fighter ops, with him.

Bikle initiated the discussion by describing the basic problem of PIOs during towing operations. He indicated that it was an infrequent problem in glider operations. It usually occurred with new pilots and was the cause of a number of fatalities over the last thirty years. New pilots tended to be overzealous in their response to changes in position behind the tow plane; they tried too hard to maintain a perfect position. The more seasoned pilots would ignore minor position changes due to turbulence or maneuvering and allow themselves to be dragged back into position rather than trying to fly back into position. Bikle explained that the M2-F1 was particularly sensitive to a PIO on tow because of its unconventional aerodynamic characteristics.

I defended Gentry even though I knew he was an extremely aggressive pilot. That wasn't the best characteristic to have in a potential PIO situation. The checkout flight in the sailplane at Tehachapi the year before had been a real demonstration of his aggressiveness. I was disappointed, but not surprised, when he precipitated his second PIO in the M2-F1.

In discussing Gentry's future with the general, Bikle decided that we would not attempt to check Gentry out in the M2-F1 again. In fact, Bikle decided we would ground the M2-F1. It had served its purpose, and there was no need to risk an accident for something that was not germane to the new heavyweight lifting bodies. We would not be towing the new lifting bodies; they would be air-launched from a B-52. We had to devote our attention to the new lifting bodies and demonstrate their capabilities. There was some experience to be gained by flying the M2-F1 in preparation for the M2-F2 and HL-10, but was the risk worth it? Bikle decided it wasn't.

With that decision, the general backed off from his opposition to Gentry's continuation in the program. He suggested a compromise. He would assign Colonel Sorlie to the program temporarily to fly the M2-F2 and then rely on Sorlie's recommendation to reassign Gentry to the program. The proposal was agreed to by Bikle.

That ended the flight history of the M2-F1. The official grounding date was August 18, 1966. Because of the fragmented records, it is not clear how many air-towed flights were made. The counts range from 71 to 90. The number of ground tows was somewhere around 400.

We had learned a lot by building and flying the M2-F1. The most dramatic benefit was the media and public visibility gained by actually flying a wingless shape and landing it like an airplane. Never mind that it was terribly inefficient as a glider. It offered the potential of a land recovery of future spacecraft on a runway just like a normal airplane. This positive vision sold NASA headquarters managers on a follow-on heavyweight program to demonstrate the higher-speed capabilities of lifting bodies.

The M2-F1 was a real success.

6

JUSTIFICATION AND DEVELOPMENT OF THE HEAVYWEIGHT LIFTING BODIES

The M2-F1 was intended to demonstrate that a lifting-body aircraft could be built and flown successfully. In the early sixties, the idea of an airplane without wings invoked the same reaction among many engineers that the idea of a plane without propellers had in the early forties. The lightweight lifting body was far from an operational vehicle, however. The M2-F1 had a wing loading a fifth that of an orbital lifting body. Its handling characteristics were, as a result, very different from those of a heavyweight vehicle. Even with the same lift-to-drag ratio, the M2-F1 was much more demanding to land than a heavyweight lifting body. Although the M2-F1's landing speed was slower, the time between the start of the flare and touchdown was much shorter than that for a heavyweight vehicle.

In early 1963, before the M2-F1 had even flown, Dale Reed was looking at an air-launched, heavyweight, rocket-propelled lifting body able to reach speeds of Mach 2. Called Configuration 2, one of its justifications was to provide data on the transonic and supersonic behavior of a lifting body. The most difficult problems for any aircraft were at speeds between Mach 0.8 and Mach 1.2, a transitional zone with mixed subsonic and supersonic airflow. At speeds above Mach 1.2, the airflow smooths out, and below 0.8 the simplified noncompressibility assumptions apply. A heavyweight lifting body would also provide data on piloting problems and workload, and on how to make the approach and landing of such a vehicle. To keep costs down, the heavyweight lifting body would use as much off-the-shelf equipment as possible.

The success of the early M2-F1 flights encouraged Paul Bikle to propose the construction of an operational vehicle. The initial proposal to NASA headquarters was to build two heavyweight M-2 vehicles, with the second vehicle being a backup in case we lost the first vehicle in an accident. X-type aircraft were typically procured with one or more backups; experience with early experimental aircraft had demonstrated the wisdom of that approach. Debris from crashed experimental aircraft littered the desert surrounding Edwards.

Headquarters personnel agreed with the need for a backup vehicle, but they proposed building an HL-10 configuration rather than a second M-2. The HL-10, a lifting-body configuration developed at the Langley Research Center, was a more refined configuration with better subsonic performance and potentially better landing characteristics. Bikle and his team were somewhat reluctant to have a different configuration as backup, since few of the structural parts would be interchangeable, but with headquarters controlling the purse strings, he had little to barter. The heavyweight program could not be built in the shoestring, semicovert way the M2-F1 had been. He agreed to an HL-10 backup vehicle.

There were some advantages to this plan. We made our benefactors happy, we obtained significant support for the program from Langley, the mother aeronautics center of NASA, and we gained a more desirable configuration with better flying characteristics. We could in this fashion compare a range of performance to define the minimal acceptable level. It was ultimately concluded at the completion of the flight program that the HL-10 had close to the optimum landing performance. A subsonic lift-to-drag ratio of approximately 4 to 1 appeared to be the optimum for unpowered landings. The lift-to-drag ratio of the M-2 was roughly 3 to 1, which was determined to be the minimum acceptable performance. It was unforgiving if you failed to fly it precisely, as we learned in a subsequent catastrophic landing accident.

Interestingly, the NASA headquarters managers were hedging their approval to build these new vehicles. They insisted that the aircraft be defined as wind tunnel models potentially capable of flight. They were not ready to commit themselves to a flight program until we gained more experience with the lightweight vehicle and obtained some promising wind tunnel results with the heavyweight vehicles. They also refused to provide funding for a propulsion system to allow the vehicles to be flown to supersonic speeds. Again, they were being extremely conservative. They obviously had cause to be conservative, since Bikle had built the M2-F1 without any formal headquarters approval. This time headquarters managers were going to keep a tight rein on the program. They were going to set the pace of the program, or at least they thought they were.

CONCEPTUAL DESIGN OF THE M2-F2 AND HL-10

Once we received headquarters approval, we spent the next several months developing a conceptual design of the vehicles and their systems. We completely ignored the headquarters direction to develop two wind tunnel models that might ultimately be flown. We designed them from scratch as manned flight vehicles capable of supersonic flight. The design specifications we developed were:

Cockpit pressurization using a high-pressure mixture of nitrogen and oxygen, rather than bottled air, to minimize the amount of water and thus prevent frosting of the windows and instruments.

Dual hydraulic systems for flight controls, keeping one system completely isolated with its own pump and battery, and the other system including ram air turbine (RAT) and primary or backup battery.

An electrical system with three primary batteries: one to drive only the first hydraulic pump; one to drive the second hydraulic system, the cockpit and window heaters, and the radios; and another to power the instrumentation system.

A stability augmentation system initially single channel only, in all three axes; it was later changed to dual pitch with manual selection.

A throttleable landing rocket, with two 400-pound thrust Lunar Landing Research Vehicle lift rockets and a 96-gallon peroxide tank from the X-15, which could almost maintain level flight with full thrust.

Landing gear—consisting of T-38 main gear, T-39 nose gear with dual corotating wheels, and no nosewheel steering—pneumatically deployed and locked in less than two seconds.

Once the specifications were completed in February 1964, we solicited proposals from the aircraft industry to build these unique vehicles. We knew we could not afford to have the contractor build them to the stringent military specifications, so we developed our own for most major assemblies and components. We did specify mil-spec (military specifications) for hardware such as nuts and bolts, tube fittings, O-rings, and seals. For the major portion of the vehicle, we recommended that the contractors use their own good design practices and good engineering judgment and increase the design margins where necessary to achieve a safe, flight-worthy vehicle. We were not critical weightwise.

It's amazing what quality of work an organization can obtain when it shows some respect for the contractors' skill and craftsmanship. Their reputation was now on the line. They weren't about to do sloppy work.

Out of the twenty-six aerospace companies that received requests from us, three came back with proposals: one each from Northrop, North American Aviation, and United Technologies Corporation. United Technologies was not a recognized airframe contractor, and we were surprised to receive its proposal, but we included it in our evaluation since we were looking for a very low-cost approach similar to the development techniques of the Lockheed Skunk Works. We had in fact recommended a Skunk Works approach and the maximum use of existing off-the-shelf components. We were not trying to develop an optimized spacecraft. It was only to be a feasibility demonstrator. The sophisticated spacecraft would come much later, or maybe never if the demonstrator results were unsatisfactory.

Each of the proposals was good, but the United Technologies proposal suffered from a lack of airframe design experience. The two other proposals were quite competitive. Northrop's was quite good both technically and financially. Northrop was extremely interested in building these vehicles; it had no space expertise and wanted to get a foot in the door.

Richard Horner, a longtime friend and coworker of Paul Bikle's, had recently become executive vice president at Northrop following time at NASA headquarters. While at NASA, he had recommended Bikle as a replacement for Walt Williams at Dryden when Williams moved to join the Space Task Group to develop the Mercury program. Horner was a strong proponent of the Northrop proposal.

We finally selected Northrop to build the M2-F2 and HL-10. The agreed fixed price was $1.8 million for the two vehicles configured for flight. NASA subsequently added half a million to include a rocket propulsion system for each vehicle. NASA provided the rocket engines, XLR-11 engines that initially had been used in the X-1 and D-558-II research aircraft. The engines were reclaimed from various museums and delivered to Reaction Motors for refurbishment and upgrading to increase thrust, from 1,500 pounds per each of the four barrels (6,000 pounds per engine) to 2,000 pounds per barrel, or 8,000 pounds total.

During the development and construction of the vehicles, Northrop invested $1.8 million of its own money to complete the vehicles. The government received two high-quality research vehicles for slightly over $1 million each at a total joint cost of $4.1 million. One industry estimate was that each vehicle would cost $15 million. The heavyweight lifting-body program was an outstanding bargain and a tribute to Paul Bikle and Richard Horner of Northrop.

The development program was an outstanding example of an innovative joint management technique. Northrop created a development team of highly experienced design engineers, experimental technicians, and mechanics who worked directly for a company vice president. The team was provided a controlled work area and dedicated shop support. NASA provided an on-site manager, designated vehicle crew members, and aircraft quality assurance and inspection personnel. NASA engineers and myself, representing the pilots, were frequent visitors and were continuously available through phone communications. With this organization, any questions, problems, modifications, or arguments about the intent of design specifications could be addressed and usually answered within hours. The development team of Northrop and NASA personnel worked extremely well. As the designated pilot for the first vehicle, the M2-F2, I was involved in a major portion of the real-time decisions. I was so involved that the team started referring to our design specifications as "Milt specs" rather than "mil-specs." This development organization was created to minimize overall development costs. It succeeded admirably.

Paul Bikle was determined to keep the cost down. His credibility was on the line. One of his techniques for tight cost control was to challenge the need for almost every component in the vehicle. For example, he decided that I didn't need an attitude indicator for the unpowered flight phase. I argued with him. I tried to convince him that I needed it for accurate flight path control throughout the gliding descent. He wouldn't buy that argument, since we would be flying only in clear weather and, if necessary, I could put some marks on the canopy for reference. His position was a little extreme, but he was trying to drive a point home. An attitude indicator was a relatively expensive instrument. He wasn't going to buy one unless we absolutely had to have it. The glide flights in the M2-F2 were made with a big hole in the instrument panel for a future attitude indicator.

C-47 DUAL ENGINE FAILURE

During the development and construction of the M2-F2, I made numerous trips to Northrop at Hawthorne, California, to review the construction progress and examine the hardware. On one trip, I flew down in our C-47, with Vic Horton sitting in as copilot. Horton was not a qualified pilot, but he had some previous pilot training. We had a number of NASA engineers and technicians as passengers who were also supporting the development effort.

The runway at Hawthorne was a relatively short runway in a densely populated area in the south part of Los Angeles. The city had surrounded it after World War II, and as a result it was no longer used as a test site for production military aircraft built by Northrop. It was completely adequate for our C-47, but there wasn't a lot of room for error with a full load. The landing was uneventful.

It was hot as we prepared for takeoff that afternoon. The takeoff was normal, but as I climbed through 500 feet, after retracting the flaps and landing gear, both engines quit completely. They didn't sputter, cough, or slowly lose power. They just flat quit running, as though someone had turned off the magnetos. I was stunned. I had never witnessed anything like it in my eight years of experience with the C-47, and I had never heard of any similar incident.

My first reaction was to check the magneto switch. It was in the proper position. I cycled it off and back on in an attempt to restart the engines. Nothing. No luck. I then reduced the throttle setting, and within a few seconds, the engines started up again. Although it seemed as though the engines were not running for several minutes, it was actually more like 5 or 6 seconds—but 5 or 6 seconds can be an awfully long time.

I continued my climb as I regained airspeed, but as I climbed through 800 feet, both engines quit again in the same manner. This time it was not just for a few seconds. I had to dump the nose over to maintain airspeed and began looking for a place to set it down. There was not an open area in sight. Nothing but buildings, streets, and a freeway filled with cars. The freeway, as busy as it was, appeared to be the only survivable landing location if I could avoid overpasses. I headed for the freeway.

As I descended through 600 feet, the engines abruptly started again. I had been frantically checking the fuel selector, mixture controls, and other critical switches and levers as I maneuvered toward the freeway, but had no luck starting the engines. All of a sudden, they seemed to start on their own initiative. This time I headed for the ocean. As an old Navy pilot, I preferred ditching rather than crash landing. The engines continued to run as I passed over the shoreline and turned north to parallel the coastline. By that time I decided to head for Edwards rather than attempting to land in the Los Angeles area. I didn't want to lose both engines on a landing approach over a bunch of houses. During the flight to Edwards, I made certain that I had an emergency landing site within gliding distance at all times. The engines ran smoothly from the time we left the Los Angeles area until we landed at Edwards. Maybe the engines just didn't like the LA smog.

We spent many hours troubleshooting the problem. We strongly suspected the mag switch unit, but we could never duplicate the problem or detect any unusual switching or shorting tendency. We finally gave up troubleshooting and began replacing any components that could precipitate such a failure. I really lost confidence in that airplane following that incident. Two engines aren't much safer than one if there is a common failure point.

M2-F2 VEHICLE CHECKOUT

The two vehicles were constructed in sequence, the M2-F2 first and the HL-10 second. The M2-F2 was completed and delivered on June 15, 1965. The HL-10 was delivered on January 18, 1966. Each was moved to Dryden, and we began vehicle checkout in preparation for the first glide flights.

The checkout of a brand-new research aircraft is a lengthy and tedious process. It begins as the vehicle is being assembled. Individual components are checked for proper operation prior to installation in the vehicle. These tests can be accomplished by the component manufacturer or by the manufacturer of the vehicle.

Once the vehicle is assembled, subsystem and system tests are conducted to ensure that each particular system works as specified or designed. In the case of a flight control system, there are several subsystems, such as the hydraulic system that powers the actuators that physically move the control surfaces. Another subsystem is the mechanical system, which moves the hydraulic actuator metering system valve to cause the actuator to extend or retract in response to the pilots' inputs. A third subsystem of the flight control system is the stability augmentation system, which provides artificial aircraft stability to eliminate any high-frequency aircraft motions.

Each subsystem is checked to ensure that it is working as specified before the combined system is checked. The same procedure is followed for each system and subsystem in the aircraft. Once these individual system checks have been successfully conducted, integrated system checks are initiated. Integrated or combined system tests are conducted to determine if there are any interactions between the systems that create problems or cause failures.

The testing we did before the M2-F2 was delivered was intentionally limited to speed up the delivery date. Bikle had an unwritten understanding with Northrop that the company would fix, modify, or redesign as necessary to ensure that the vehicle and its systems would function as intended. We did the detailed system testing after we received the vehicles, viewing it as an opportunity to gain experience working with the systems. This testing, modification, and retesting took a year before we decided we were ready for the first M2-F2 glide flight. Again, we were being extremely conservative. We did, however, encounter some significant problems during our testing.

The crew was testing the flight control system one day when Bikle and I wandered down to observe the daily activity. The flight control system was powered up with the stability augmentation system engaged. During a lull in the testing, someone tossed a shot bag on the upper pitch flap, which started to oscillate up and down in a cyclic manner. It continued to oscillate, rapidly increasing in amplitude, until the stability augmentation system was deactivated.

It took some time to solve that problem. It was the result of the location of the system's rate gyros, which were mounted on the same bulkhead as the flap actuator. When the shot bag was placed on the flap, it created a load on the actuator that resulted in a deflection of the mounting bulkhead. The rate gyro sensed this deflection and assumed it was a vehicle motion that should be damped. The result was a self-sustaining control system oscillation that could have been dangerous if it was encountered in flight. The problem was relatively easy to fix once it was identified.

Another problem was a little tougher to fix. I noticed during a combined systems test that when I depressed the radio mike switch, the pitch stability augmentation system disengaged. This was obviously unacceptable. The vehicle required pitch damping during several critical mission phases such as launch. It took a while to isolate the problem. We had to do some extensive electrical tests to solve it. Even after we solved the problem, we weren't sure that there weren't additional problems lurking in the electrical system.

With the successful completion of the combined systems tests on the M2-F2, the next step was to check the combined systems while the vehicle was mated to the B-52 mothership. Prior to all these tests, all systems on the B-52 that interfaced with the M2-F2 had undergone their own series of subsystem, system, and combined systems testing. During the mated combined systems tests, when the M2-F2 was attached to the B-52's launch pylon, all systems in both aircraft were activated and a simulated mission was conducted. The control-room personnel monitored all telemetry data and initiated the checklist actions as the simulated mission progressed.

All of the personnel involved in an actual flight were manning their designated positions in each aircraft, the control room, and the various ground vehicles that participated in the pre- and postflight operations. It was not unusual to encounter numerous problems or anomalies during these tests. Many of the problems could be worked out in real time, since the operation was not limited by fuel or flight time. When a problem did show up, those directly involved would work to solve the problem, while everyone else had to wait until they were done and the test could be reinitiated. For many of the ground crew, this waiting for someone else was the worst part. This type of test could drag on for hours or even days in cases where the entire test had to be rerun.

One problem area was the launch pylon adapter. An adapter from the hooks in the B-52 wing pylon to the M2-F2 was required to compensate for the shorter length of the M2-F2 and to position the M2-F2 cockpit ahead of the B-52 wing leading edge. This was essential to enable the M2-F2 pilot to eject while still attached to the B-52, a feature carried over from the X-15 for pilot safety in the event of a catastrophic problem while still mated to the B-52.

Northrop built the adapter using standard low-strength construction steel typically used in large commercial buildings or bridges. Weight was not a problem, so we routinely overdesigned various vehicle and mothership components to add margins of safety. The Northrop-built adapter was way overdesigned, but our welder noticed some small cracks in the welds and insisted on redoing them all. Our welder was exceptionally capable. He had worked for NASA for many years, working with all kinds of exotic materials on the various research aircraft. He was not going to allow me to fly the M2-F2 until he had reworked all the welds on that adapter. It took him about a week to complete the task, and then we got serious about flying the M2-F2.

CAPTIVE FLIGHT TESTS OF THE M2-F2

Before we made the first glide flight, we had to test the structural integrity of the pylon adapter and the ability of the M2-F2's systems to operate under conditions of flight. A captive flight could reveal some environmental problems, such as canopy fogging or icing. Other potential system problems were due to the cold soaking that occurs at launch altitude, where the outside air temperature is normally around minus 65 degrees F. As with a car exposed to extreme temperatures in a cold climate, some things just don't work under those conditions.

The effects of cold soaking can be catastrophic in an aircraft that is landing without power. If the landing gear fails to deploy or deploys slowly because of cold soaking, the vehicle may hit the ground without landing gear, which can be catastrophic at landing speeds approaching 250 miles per hour. Hydraulic systems tend to respond more slowly at extremely low temperatures, which could precipitate control problems due to lags in control response. These are the kinds of problems that can't be discovered in the snug, controlled environment of a hangar but that test teams hope will surface during a captive flight. A glide flight is less than 4 minutes in duration, and the vehicle doesn't get a chance to warm up during the rapid descent to a landing.

I decided that I wanted to witness a demonstration of the critical integrity of all the elements involved in securing the M2-F2 to the B-52—the B-52 pylon, the adapter from the B-52 hooks to the M2-F2 hooks, and the M2-F2 vehicle itself—before I climbed into the M2-F2. This test involved two short-duration captive flights with the unmanned M2-F2 attached to the B-52. The flights were planned to utilize the long runway on the south lake bed. Each involved a takeoff, then a short stabilized flight at 100 feet altitude at 220 knots, and then a landing at the south end of the runway. A speed of 220 knots was selected since it was our planned climb speed on our first manned captive flight.

The two unmanned flights, M-C-1 and M-C-2 ("M" for M2-F2, "C" for captive), were both made on October 21, 1965. The M2-F2 stayed attached to the B-52 at takeoff and landing approach speeds. Everything went smoothly, and we were ready for our next series of tests.

The next test was a taxi test to determine how well the vehicle handled during simulated landings. I used the landing rockets to accelerate up to speed, and then, after shutting the rockets down, I made gentle turns to evaluate the steering capability. During these taxi tests, I accelerated the M2-F2 up to speeds in excess of 100 knots. The results of these tests revealed that the M2-F2's ground handling characteristics were poor. I found it was hard to turn the vehicle during some runs, although on others it was hard for me to stop it from turning. There was no nosewheel steering, so I had to rely on asymmetric braking to turn the vehicle. The brakes were T-38 brakes, not known for their effectiveness, and as a result, I had a major problem steering the vehicle.

We determined that the vehicle was sensitive to differential shock strut height of the main gear. Minor variations in height due either to servicing or leaks resulted in strong turning moments that could overpower the poor brakes. We attempted to control the checkout and servicing of each main gear strut rigidly, but occasionally we still had problems during landing rollout.

With the pylon qualified and the taxi tests complete, we were now ready for the manned captive flights. The first manned captive flight, M-C-3, made on March 23, 1966, was originally intended to be a short flight around the landing pattern to demonstrate successful deployment of the landing gear. The flight plan was subsequently modified to include testing to verify the structural integrity of the M2-F2, the adapter, and the B-52 pylon at increasing speeds with the upper and lower control flaps of the M2-F2 at different positions.

During the flight we used various control-system pulses by the B-52 pilot to excite the combination of pylon, adapter, and M2-F2 and verify good structural damping of the ensuing wing and pylon motions. A rudder kick by the B-52 pilot seemed the best maneuver to excite the pylon motions. I began to get concerned on some of those rudder kicks. Sitting in the M2-F2 cockpit, I felt as though the M2-F2 would be flung off the wing of the B-52. As it turned out, however, the pylon and adapter were strong enough to withstand the worst rudder kick, and we successfully concluded that series of tests.

We also checked the peroxide jettison system, the airspeed system, the hydraulic system ram air turbine (or RAT, which provided emergency hydraulic pressure), and the landing gear extension after a rapid descent from altitude. The RAT did not function properly. It deployed successfully, but it did not generate any hydraulic pressure. The landing gear worked as designed. Some canopy icing problems were noted; they required some modifications to ensure that I had adequate visibility for landing. The same problem was encountered on an early X-1 flight, requiring that the chase pilot talk the X-1 pilot down to a safe landing. That was a hairy operation, one that I wouldn't want to be involved in.

A captive flight, if properly planned, could provide as much or more data than a free flight. The M2-F2 and B-52 systems could, for example, provide a simulation of the aerodynamic loads on the mated M2-F2 and potentially verify the predicted vehicle motions after launch. The M2-F2 data system could provide control-surface loads and hinge moments at various speeds and deflections to verify wind tunnel predictions. The M2-F2 airspeed, altimeter, and rate-of-climb pressure instruments could be partially calibrated against the B-52 instruments during mated flight. After a well-planned captive flight, there were not too many unknowns about a research vehicle, except for its free-flight flying qualities.

The next flight, M-C-4, flown on May 6, 1966, was a complete simulated launch flight. On this flight we not only checked out each system, but we also verified the checklists that we had prepared for the first free flight. On captive flights of this type, the standard checklist is used up until launch time. At that time, an "Abort" is declared, and we then proceed to accomplish some other objectives to obtain data and experience for the free flight. On this flight, we attempted to simulate the free flight of the M2-F2 by making a rapid descent while mated to the B-52. We expected the maneuver to reveal any problems due to the high-altitude cold soaking.

One of the most critical concerns was a successful landing gear deployment after the rapid descent. The landing gear deployed as designed, which took a big load off my mind. We deployed the RAT again on this flight, and it worked perfectly. During the descent, I had a stability augmentation system failure and some cockpit icing. After landing, I had a tough time opening the canopy. It took a lot of force on the canopy release handle to unlock the canopy.

The next captive flight, M-C-5, on June 7, 1966, was a repeat of M-C-4. After working out the problems that were encountered on flight M-C-4, it was uneventful. Everything seemed to work. We had some minor anomalies, but nothing that required any major modifications or rework. We believed we were ready for a free flight. A tragedy the next day forced a reassessment.

JOE WALKER'S DEATH

Two of Joe Walker's children were to graduate from school in Lancaster on June 8, 1966. During that same day, Walker was scheduled to participate in a formation of aircraft that all utilized General Electric jet engines. The formation aircraft included the XB-70, an F-4, a YF-5, a T-38, and Joe Walker in a NASA F-104.

This formation flight was conceived by the General Electric Company to obtain in-flight pictures of the aircraft for advertising purposes. Clay Lacy, a well-known professional aerial photographer, was hired by GE to photograph the formation. GE had to gain approval for this flight from the Air Force Flight Test Center and from NASA to obtain the various aircraft. The Flight Test Center was asked to provide the XB-70 and the other aircraft, except the F-104.

NASA was asked to provide the F-104. This type of mission was officially frowned on by Air Force headquarters, but it was occasionally approved to support the aerospace contractors. To minimize any potential criticism, the photographic mission was added to a scheduled test mission. This would eliminate any accusation that a special flight was flown to obtain publicity pictures for an aerospace contractor.

The General Electric request for NASA participation was directed to Joe Walker, who was NASA Dryden's chief pilot. Walker was in favor of participating, but he had to get Paul Bikle's approval. Bikle was not enthused about the idea. He finally told Walker that he could do it, but it would have to be done on his own time and not on official duty time.

On the day of the scheduled flight, Walker took off in one of NASA's new F-104N aircraft, N813NA. NASA had just recently procured three new F-104N aircraft to use as proficiency and support aircraft for the X-15 program. Walker rendezvoused with the other four formation aircraft and the Lear Jet photo aircraft just north of Edwards. The XB-70A was being flown by Al White, the North American test pilot, and Maj. Carl C. Cross, a new Air Force Flight Test Center project pilot. This was his first XB-70A flight. The YF-5A was being flown by John Fritz, a GE test pilot; the T-38 was being flown by Peter Hoag and Joe Cotton; and the Navy F-4B by a crew from Point Mugu. The formation, consisting of the T-38 and the Navy F-4B off the XB-70's left wing and the F-104N and YF-5A off the right, flew a racetrack pattern.

Without warning, Joe Walker's F-104 suddenly pitched up and rolled to the left, toward the XB-70. It continued up and over the XB-70's wing, colliding with the twin vertical tails, severely damaging the right tail and tearing off the left tail. The F-104 broke in half and exploded in a ball of flame following the collision and burned as it fell toward the desert floor. Walker did not eject.

Joe Cotton in the T-38 immediately called, "Midair! Midair!" and then, "You got the verticals, this is Cotton, you got the verticals—came off left and right." Aboard the XB-70, neither White nor Cross had felt the collision, and White thought that two of the other planes had collided. After 16 seconds of stable flight, the XB-70 suddenly rolled off to the right and seemed to enter a spin. It spun down and crashed in the desert about 12 miles north of Barstow, California. Al White managed to eject successfully after encountering some major problems with his escape capsule. Major Cross did not eject.

I was in the pilots' office when the call came from base operations that a midair had occurred involving the XB-70 and our F-104. It was an extremely tense period of time until we received confirmation that Walker was still in the cockpit after the airplane crashed.

Joe Vensel, NASA's chief of flight operations, and Paul Bikle departed almost immediately to drive into Lancaster to inform Walker's wife, Grace, of the accident. I was instructed to take a photographer and the F-104 crew chief out to the crash site to obtain any available evidence to support the accident investigation board.

It was a long and somber drive out to the accident site. The majority of the personnel accumulated at the accident site were primarily interested in the XB-70. We finally located the F-104 northwest of the XB-70. The F-104 had broken into two major pieces. The engine section and most of the wings came down about 2 miles northwest of the XB-70, while the cockpit section fell half a mile farther northwest. The cockpit was shattered and burned. Walker was still strapped firmly in his seat. He never had a chance to eject.

Walker's family did not make it to the graduation ceremonies that evening. His death was a devastating blow to his family and to his coworkers. Whenever a pilot was killed, like Walker or Mike Adams the following year, my family closed in together—we all became thoughtful and quiet.

No one had a plausible reason for the accident. Walker was one of the foremost test pilots in the country and possibly the world, and yet the F-104 collided with the XB-70 and not vice versa. There were many theories on the cause of the accident but no conclusive answer. One theory blamed strong local vortices near the wingtip that sucked the F-104 into the XB-70. The accident investigation board finally concluded that Walker misjudged the position of his horizontal tail relative to the downward deflected tip of the XB-70 wing. As he flew in formation, the F-104's tail touched the downward deflected tip of the XB-70 wing, which then caused his airplane to pitch up and roll over the wing and hit the two vertical tails.

This theory is somewhat more believable because of the extreme difficulty in judging relative position on a large deltawing airplane. The downward deflected wingtip of the XB-70 created an added complication in judging relative separation or clearances between the two aircraft. Close formation flying on the XB-70 ahead of the wingtip was an accident waiting to happen.

I went by the Edwards hospital that evening after returning from the crash site. Al White had been flown there by helicopter. He was awake and talking when I arrived, but he was in terrible pain. The impact bag on the bottom of his escape capsule had failed to inflate properly. The capsule, with White inside, hit the ground hard, hard enough to fracture a vertebra.

He was also suffering from a badly bruised right elbow. When he pulled up on the handle to eject, his elbows stuck out in the path of the capsule's clamshell doors. When the clamshell doors failed to close, the ejection sequence stopped. His arms were immobilized by the extreme pressure of the doors, yet he had to free them or his capsule would remain in the aircraft. He finally freed his left arm and then used that arm to pull his right arm free, leaving some skin and meat on the outer edge of the clamshell door. Once the doors closed, the capsule ejected. After ejection, the parachute opened normally, but the capsule impact bag malfunctioned.

When White was questioned about Major Cross, he could only surmise that Cross had encountered similar problems with the capsule's doors. Cross was somewhat larger than White and would probably have had even more trouble squeezing himself into the capsule to ensure that the clamshell doors closed properly. Ejection capsules were attractive for high-speed aircraft, but they were much more complex than an ejection seat. Their reliability suffered.

NASA subsequently borrowed the remaining XB-70 from the Air Force to conduct a flight research program. The reliability of the escape capsule became an issue. I was a member of a board that evaluated the capsule reliability and potential improvements. The proposed improvements offered some increases in reliability, but the cost to implement them was prohibitive considering the NASA dollars available. Realizing that modifying the capsule and requalifying it would cost millions and take years (it would require rocket sled tests to simulate Mach 3 speeds), Bikle came up with a more pragmatic suggestion. He was concerned about pilot safety, but he also had the cynical attitude that test-pilot concerns about safety depended on how much bonus money they were getting. Bikle's proposal was to wrap a $100 bill around the XB-70 stick rather than trying to modify the capsule. That would have been substantially cheaper.

THE FINAL CAPTIVE FLIGHTS

Joe Walker's death gave us all pause and caused us to rethink the M2-F2 schedule. We had been about ready to make the first free flight before Walker was killed. John McTigue (called Tiger John), the lifting-body program manager, wasn't too comfortable with the status of the stability augmentation system in the M2-F2. We had had some unexplained dropouts or systems failures during previous testing. We had made some changes after each anomaly, but we weren't positive that we had corrected the real problem. It was a randomly appearing problem, and it did not always exhibit the same symptoms.

Following Walker's death, McTigue directed the M2-F2 crew to reexamine all the electrical wiring and associated components to ensure that we didn't have any other potential problems in the system. The requalification tests were redone and modifications were made whenever they appeared to improve the reliability of the system. This effort delayed the program for about three weeks. I didn't mind waiting; I just didn't want to run into any surprises in flight every time I pressed the mike switch.

The next flight, M-C-6, was flown on July 6, 1966. The flight was a planned captive flight specifically designed to evaluate thoroughly the stability augmentation system after the inspection and modifications. We had some minor problems during the flight. The landing rocket system did not work properly using the throttle, although it worked using the throttle bypass mode. We had some radio problems, but nothing that would preclude a free flight. The chase pilot noted some vibration of the upper flap prior to pressurizing the hydraulic system, and I noted some canopy icing problems that still persisted after we had modified the deicing system. The icing would not have compromised a flight, however. Everything worked fine, including the stability augmentation system that had been of concern to McTigue.

A free flight was scheduled for July 11, 1966. During the systems checkout prior to launch, we encountered trouble with the landing rocket when attempting to fire it using the throttle. We aborted the flight and worked on the landing rocket system the remainder of that day. The system was modified, checked, rechecked, and finally declared ready for flight.

The free flight was rescheduled for the following day.

7
"LOOK MA, NO WINGS"

In laying out the flight plans for the early M2-F2 flights, it was assumed that I would make the first few flights before beginning the checkout of the other pilots. This would allow me to gain some experience that would enhance my ability to cope with any problem that might be encountered during the preliminary envelope expansion phase. Based on simulation, we knew there were some questionable areas of flight. It would be my job to fly into each of those areas to determine whether it would be safe to allow the other pilots to venture into them.

The primary objective of the first free flight was to get the vehicle on the ground safely on the intended runway. The overall flight plan was tailored to concentrate on verifying the landing technique that we had developed on the simulator and in the F-104 support aircraft. We had some additional objectives for the first flight, but they were secondary, and if any problems were encountered on the flight, they would be abandoned.

One secondary objective was to verify that we could safely fly into the predicted PIO (pilot-induced oscillation) region. Others were to verify that all of the vehicle's subsystems worked properly in an integrated fashion and to validate the predicted launch dynamics. It was difficult to predict what the dynamics might be as the vehicle dropped away from the B-52. Vehicle launches off a mothership were not always without incident. Previous rocket planes had occasionally rolled or pitched during launch, raising concerns about a recontact or collision after launch. This recontact problem was not uncommon when dropping fuel tanks or bombs from aircraft; they have collided with the launch aircraft on numerous occasions. It was not an easy problem to analyze. We selected the most benign launch conditions that we could find on our simulator for the first flight.

M2-F2 FIRST FLIGHT PLAN

The first flight plan called for a launch at 45,000 feet at 165 knots indicated airspeed. Immediately after launch, I was to increase airspeed to 220 knots prior to beginning a left turn at 39,000 feet altitude roughly 26 seconds after launch. I was to maintain 220 knots during the turn and roll out after turning roughly 90 degrees parallel to the north shore of the lake bed. On reaching an altitude of 30,000 feet at 59 seconds after launch, I was to push over to a minus-3-degree angle of attack to pick up 300 knots indicated airspeed.

At 22,000 feet, roughly 80 seconds after launch, I was to initiate a flare using 1.6 g to come level at about 18,000 feet altitude. As I decelerated in level flight through 200 knots, I was to fire the landing rocket in a short burst to verify that it was working properly and that the thrust was aligned properly to minimize any transient vehicle motions. After the rocket burst, I was to push over to a zero angle of attack and build up airspeed to 190 knots.

On reaching 16,000 feet altitude, I was to roll into a 45-degree bank to turn onto final approach to Runway 18 on the north lake bed. During and after the turn, I was to accelerate to 300 knots to be prepared for the landing flare, which would begin at roughly 3,300 feet altitude, or 1,000 feet above the lake bed. Touchdown should occur approximately 3.5 minutes after launch. There was a lot of thought that went into defining that flight plan. The ground track involved a short downwind leg, a 90-degree left turn to a long crosswind leg, and then another 90-degree left turn to a short final leg to land on Runway 18. The first turn shortly after launch was intended to allow me to evaluate the turn capability of the vehicle and its lateral-directional control characteristics. I would find that out quickly after launch and be able either to modify the flight plan or to eject if the vehicle was not adequately controllable. It's nice to find that out early in the flight, since I had only about 3.5 minutes to learn to fly the vehicle before I had to land it.

Following the first turn, the plan called for a practice flare at altitude to determine whether the vehicle had sufficient lift capability to arrest the rate of descent completely. Again, we were trying to find out as quickly as possible whether the vehicle would be controllable through the critical phases of the flight. If the practice flare at altitude worked out as planned, I would be reasonably confident that the final flare could be safely accomplished. If I had problems during the practice flare, I again could either modify the flight plan or eject at a safe altitude rather than just above the lake bed, which might be the case if I had not tried the flare at altitude.

The practice flare at altitude also gave me an opportunity to evaluate the predicted lateral-directional PIO problem, which we had observed during our simulation. The PIO problem was predicted to be more severe at low angles of attack or high airspeed; however, we had to accelerate to high airspeed prior to flare to have enough energy to arrest the rate of descent and come level for touchdown. We were thus committed to enter the region of greatest concern. The practice flare at altitude allowed me to evaluate the PIO tendency high above the ground. If the PIO was more severe than predicted, I could reset the aileron-rudder interconnect to a more benign setting prior to the final flare, or I could use a lower airspeed prior to flare. The lower airspeed would alleviate the PIO problem, but it would also reduce the time available to complete the flare. We were boxed in on available options. As a last resort, I still had the option to eject—not a desirable option, but still better than the final option.

In a vehicle as unconventional as the M2-F2, the pilot pretty much defines the first flight. Everyone knows that it is his butt that is on the line. There is no one with any pertinent experience to guide him in developing a flight plan. There are people who can advise him based on wind tunnel tests, analysis, and simulation, but the pilot has to weigh the overall odds of success and failure and select the plan that makes him feel the most comfortable.

The test team will always develop a set of options to cope with all the problems that can be foreseen, but if these don't work, the pilot has to have his own set of options in case the flight doesn't proceed as planned. As a result of my research flying experience, I never relied completely on ground control or the chase planes to solve every potential problem. I always had a temporary haven, a relief valve, or an escape maneuver for every conceivable problem that might arise. In the test business the pilot is always thinking of what might go wrong and what he might do to cope with that situation. That is the secret of longevity. A fitting epitaph for a test pilot is, "I didn't think of that one."

M2-F2 FIRST FLIGHT

The morning of July 12, 1966, was surprisingly cool for Edwards. I had my flight jacket and gloves on as I walked out to the B-52. The B-52 was parked on the Dryden ramp for this flight. We had no propellants onboard other than the peroxide for the landing rockets, so we didn't need to use the X-15 servicing area to load propellants. It was just after 5 A.M. when I arrived at the B-52.

The B-52 crew and the M2-F2 crew had arrived at about 0400 to begin final preparations for flight. They had loaded the hydrogen peroxide for the landing rockets, and they had filled the various gas bottles for peroxide pressurization, cabin air, and landing gear deployment. They had also installed and checked the batteries, which provided the electrical power to drive the hydraulic pumps and various other systems.

The M2-F2 was ready to go when I walked up to the B-52. Someone had sent a photographer out to take still photographs of the operation. The photographer lined up all the flight crew members of the B-52 and M2-F2 for several shots, and then we were free to climb aboard our aircraft. I was wearing a standard cloth flight suit and a jacket instead of a pressure suit. I didn't need a pressure suit on this flight because I would be coming downhill after launch; pressure suits are necessary only above 50,000 feet altitude.

I climbed up a ladder to get into the M2-F2. We didn't have a platform ladder like we used for X-15 entry. The M2-F2 was much closer to the ground since we had a large adapter between the B-52 pylon and our M2-F2 hooks. The M2-F2 hooks were only about 6 feet above the ground, which placed the M2-F2 cockpit within 5 feet of the ground. On previous captive flights, I had been surprised when the B-52 landed and lowered its nose. It seemed as though the M2-F2 would hit the runway as the B-52 nose slammed down.

Once I was in the cockpit, our personal-equipment technicians began hooking up my g-suit, parachute, bailout kit, oxygen mask, and radio. After checking the M2-F2 and B-52 breathing oxygen and radios, we began the cockpit checklist. The crew chief, Bill LePage, and our inspector, John Reeves, participated in the checklist procedures. We verified the positions of all the switches and levers in the cockpit and then called out the readings of all the instruments on the various panels. Prior to closing the canopy, we checked the radio and intercom communications with all the ground control stations. When the canopy was finally closed and all the various lines and electrical cords were disconnected from the M2-F2, the B-52 crew began their starting checklist. They began cranking engines about 10 minutes after my cockpit was closed.

The B-52 crew (Fitz Fulton as pilot, Maj. Jerry D. Bowline as copilot, Vic Horton as launch panel operator, and an Air Force enlisted man as crew chief) completed their own checklists, and we finally began taxiing just after 0600. It was a rough ride during taxi because of the deformed B-52 tires, but it tended to smooth out as the tires warmed up. Several ground vehicles followed us out to the runway, along with Chase 1 (a T-38) and the rescue helicopter. At the end of the runway, the ground crew removed the safety pins from the launch hooks, and we were ready to go. Chase 1 took off before we did to join up with the B-52 during takeoff. The B-52 finally took the runway and began its engine check. The engine check ended with all engines at full thrust; after a brief countdown Fulton released the brakes and we were on our way.

As we passed the aircraft crews in their ground vehicles at the edge of the runway, I gave them my personalized one-finger salute. The helicopter accelerated down the runway alongside us until we outran him about midfield. The takeoff was abrupt. The B-52 lifted off almost immediately after the nose was raised, much quicker than during a takeoff with the X-15. The M2-F2 was much lighter than the X-15, 6,000 pounds compared with 33,000 pounds for a loaded X-15.

After takeoff, Fulton began climbing as we circled the lake bed. The launch would be in the immediate vicinity of the lake bed, just east of the rocket site. At 15 minutes to launch, we began activating systems in the M2-F2 and checking their operation. First was the battery system, which supplied all the power to operate the other systems in the M2-F2. We did not have an auxiliary power unit like the X-15 had to supply power.

The next system activated was the hydraulic system, two independent systems that powered all the flight control surfaces. As we activated the hydraulic system, I verified that the low-pressure lights went out as the pressure built up, and then called out the stabilized system pressures. Next we checked the SAS, or stability augmentation system, to verify that all the warning lights were out. At about this time, we passed through 40,000 feet altitude on our way to 45,000 feet, our launch altitude.

We then began checking the various control-surface positions as I moved the control stick and rudder pedals to verify that we were getting proper deflections. Once we completed these tests, we set the various control surfaces at their proper positions for launch. Following that, we checked the proper operation of the SAS during control pulses initiated in the B-52. Fulton would pulse the B-52 elevators, causing the B-52 nose to oscillate up and down, and we would verify that the M2-F2 SAS was responding properly to those motions. Fulton would do the same with the B-52 rudder, causing the B-52 to yaw and roll slightly. We would again check the M2-F2 SAS to ensure that the system responded properly.

Next we checked the M2-F2 radios on our mission frequency and then on our backup frequency. I had a little trouble reading NASA 1 ground control on our primary frequency, so we switched all communications to the backup frequency. The final system to be checked was the landing rocket system. I armed the system and then opened the throttle to verify we were getting the proper thrust. The system worked as it was supposed to, and we began the final countdown.

NASA 1: Roger, and let's call this four minutes.
B-52: Roger, four.
NASA 1: Real good throttle. Shallow your turn a bit, Fitz.
B-52: Roger, shallowing that a little bit.
THOMPSON: Okay, radio check is all completed, and I'm on M-2 radio.
NASA 1: Roger.
THOMPSON: Defog going to emergency and then M-2.
NASA 1: Roger.
B-52: Forty-five thousand [feet].
HORTON: NASA 1, do you want the B-52 radar [beacon] off prior to launch?
NASA 1: Stand by.
THOMPSON: Okay, canopy defog is on M-2. Windshield defog is on M-2. Oxygen is on M-2. Going to M-2 cabin air. [I am switching to internal M2-F2 systems in preparation for launch.]
NASA 1: Okay, roll out [heading] zero zero five, Fitz.
B-52: Roger, zero zero five, we're coming through one zero five now.
THOMPSON: I got two thousand seven hundred on M-2 cabin air and one thousand eight hundred fifty on oxygen.
NASA 1: Okay, three minutes, and make that zero zero three on your heading, Fitz.
B-52: Roger, zero zero three.
THOMPSON: Cabin heat switch going to M-2.
NASA 1: Roger.
THOMPSON: Cabin heat still on high.
NASA 1: Roger, and systems are still all good here, Milt.
THOMPSON: Okay. Release low-pressure light is out.
NASA 1: Okay, let's come left to zero zero zero, Fitz.
B-52: Zero zero zero.
THOMPSON: SAS and backup lights are out.
NASA 1: Roger.
THOMPSON: And what's your time?

NASA 1: We're coming up on two minutes. Fitz, your launch heading will be zero zero five.

B-52: Roger, do you want me to come back to that at one minute?

NASA 1: That's affirmative. Two minutes now.

B-52: Two minutes now.

NASA 1: Get the B-52 and pylon cameras.

HORTON: Affirm, and I will turn the radar [beacon] off at this time.

NASA 1: Roger, radar off, and that's good, Fitz, turn right to zero zero five.

B-52: Zero zero five. Coming through one-eighty-four speed, slowing down.

NASA 1: Roger. One three one, we are one minute to drop.

THOMPSON: Okay, level your wings, Fitz, and I will uncage my gyro.

B-52: Wings are level now.

NASA 1: Okay, we'll call one minute now.

THOMPSON: Okay, gyro uncage.

NASA 1: Roger.

CHASE 1: I'm losing . . . You'll have to go ahead.

CHASE 2: Rog, Jerry.

NASA 1: Okay, cockpit camera on now.

THOMPSON: Okay, coming on.

NASA 1: All systems go, thirty seconds now, Milt.

THOMPSON: Okay.

B-52: On speed.

NASA 1: Fifteen seconds now.

THOMPSON: Roger.

NASA 1: Ten seconds now.

THOMPSON: Okay. Five, four, three, two, one, release! Okay.

The launch was surprisingly mild. I was anticipating a hard launch similar to what I was accustomed to in the X-15, but on this launch I seemed simply to fly away from the B-52. There was no violent transition from 1-g flight on the B-52 to 0-g flight in the research vehicle. There were no aggravated rolling or lurching motions. I simply began flying on my own after release from the B-52. What a pleasant surprise.

I did roll off slightly, but I was able to roll back quickly to a wings-level attitude and began setting up for the first planned turn. The hard launch in the X-15 required a second or two of mental recovery time to react after launch. This launch was a piece of cake. Part of the reason for the mild launch was the reduced trim condition postlaunch. In the X-15, the aircraft was trimmed for 0-g flight after launch. In the M2-F2, the vehicle was trimmed for 0.5 g after launch. We were comfortable with a lower rate of separation between the B-52 and M2-F2 on the basis of wind tunnel tests. The transcript continues:

NASA 1: Check your dampers.
THOMPSON: Roger, they're all on.
NASA 1: Roger. Track looks good.
CHASE: It's a beauty.
NASA 1: Heading is good.
THOMPSON: Okay.
NASA 1: Coming up on forty thousand feet now; check alpha [angle of attack] and airspeed. And, start your turn.

It was 26 seconds after launch, and I was descending through 39,000 feet altitude when I began the left turn onto base leg. The turn was roughly a 90-degree turn to parallel the north shore of the lake bed. The turn progressed without incident. The vehicle seemed to respond well in pitch as well as roll.

THOMPSON: Okay. Flying real nice.
NASA 1: Give us airspeed and alpha as soon as you roll out, Milt.
THOMPSON: Okay.
NASA 1: Turn her about an extra ten degrees left when you roll out.
THOMPSON: Okay. Pushing over.

On completion of the turn, I pushed over to begin picking up airspeed to obtain 300 knots before beginning the practice flare at altitude. Almost immediately after pushing over, I noticed a slight lateral-directional oscillation, which seemed to confirm that I was nibbling on the edge of a PIO, just as our simulation predicted. To reduce the PIO tendency, I reduced the interconnect ratio slightly. This seemed to solve the problem.

I began the practice flare early on reaching 290 knots at the proper altitude. The flare was very positive, and it was easy to maintain the 1.5-g acceleration during the flare. I came level at about 18,000 feet and continued to decelerate until I reached 200 knots. At that point I fired the landing rocket to verify that it was working properly and that the thrust was properly aligned. The rocket worked as advertised with no misalignment. I then pushed over to begin building up speed.

NASA 1: Okay, add one degree on your upper flap.
THOMPSON: Okay.

NASA 1: Okay, do you have the field in sight?

THOMPSON: Affirm.

NASA 1: Check your dampers and interconnect and start your turn anytime. Start your turn, Milt.

THOMPSON: Okay.

NASA 1: Okay, we have you going through fifteen thousand. Have you at ten thousand, Milt. Over the highway, Milt.

THOMPSON: Okay.

Right after I said, "Okay," I ran into big problems. On reaching 16,000 feet altitude, I had rolled into a left turn onto final approach. As I completed the last turn onto final approach, I had decided to change the interconnect ratio between aileron and rudder deflection to reduce any control sensitivity problems. The vehicle had been flying nicely, but I sensed that I was close to a potential controllability problem. I began reducing the aileron-rudder interconnect ratio as I rolled into the final turn and pushed over to increase airspeed. Just as I feared it might, the vehicle became very sensitive in roll. I would get more roll rate than I wanted when I deflected the stick. I immediately reduced the aileron-rudder interconnect another notch. The vehicle became even more sensitive. It began to oscillate in roll through bank angles of plus or minus 10 degrees. It almost seemed as though the interconnect ratio was hooked up backward, giving me more rather than less rudder with aileron deflection.

At this time, I was on final approach in a 27- to 30-degree dive at 300 knots. I was at 10,000 feet altitude (about 8,000 feet above the lake bed), and my rate of descent was approximately 18,000 feet per minute. I was 27 seconds away from impact if I didn't get the vehicle under control.

I reached down to grab the interconnect ratio changer and rotated it to further decrease the ratio between the rudder and aileron deflection. The rolling motions became more severe. I was rolling between plus and minus 45 degrees of bank (45 degrees left bank to 45 degrees right bank) at a rate of about 90 degrees per second. The rolling motions were becoming violent. I reached down one more time to reduce the aileron-rudder interconnection ratio and moved the control all the way to the stop. Theoretically, I should have reduced the ratio of rudder to aileron to zero. The roll motion became even more violent. The bank angles varied from 90 degrees left to 90 degrees right in less than 1 second.

The simplest and surest way of stopping a PIO is to take your hand off the control stick—no pilot input, no pilot-induced oscillation. It wouldn't do to simply stop making control inputs. In a highly charged life-and-death situation, your hand will continue to make instinctive and reactive control inputs even though your mind is telling your hand to remain motionless. I knew I was in a PIO. I knew a PIO was predicted for this flight condition, and yet, subconsciously I was making large rapid control inputs that were aggravating the oscillation.

It takes a tremendous amount of willpower to let go of the control stick when the vehicle is totally out of control and the ground is just seconds away. After all, I was a seasoned test pilot and a good one. I had extensive experience in all kinds of strange vehicles. I should be able to cope with a simple PIO. Unfortunately, many pilots have stumbled into the arms of the Grim Reaper while still trying to convince themselves that they can salvage the airplane by continuing to fight a PIO.

Let go of the stick, stupid! Let go! Let go of the stick and reassess the problem. There has to be another solution. This one isn't working. I let go of the control stick and looked down at the interconnect ratio changer. The ratio changer was positioned at its maximum value, exactly opposite to what I had intended. I had moved the ratio changer in the wrong direction. I had caused the vehicle to become overly sensitive to control inputs and then aggravated the problem by trying to fight the PIO. The ratio changer in the actual M2-F2 was different from the ratio changer in the simulator; it was a wheel in the vehicle and a lever in the simulator.

Once I saw what I had done, I immediately reset the ratio changer to its correct position. The vehicle motions had ceased once I let go of the stick, and now I tentatively applied a control input to see what would happen. The vehicle responded correctly. The sensitivity problem had disappeared. I immediately rolled back to a wings-level attitude and then turned my attention to the altimeter to see if it was time to begin the final flare.

I was rapidly approaching an indicated 3,300-foot altitude, which was my planned flare initiation altitude. I looked out ahead to see if I could see the rescue helicopter. I had requested that the helicopter hover near the south end of the runway at 3,300 feet mean sea level to provide me an altitude reference for flare initiation (pressure altimeters tend to lag at high rates of descent). When the helicopter appeared to pass through the horizon, I would be at 3,300 feet altitude, or 1,000 feet above the lake bed. I saw the helicopter just seconds before it rose up to the horizon, and then I began a slow pull-up to come level just above the lake bed. I noticed the smoke flares alongside the runway as I was easing the nose up. The surface wind appeared to be light and variable based on the smoke movement. I checked over my shoulder to see if my chase aircraft was in position, but I couldn't see him. He apparently got out of position or was forced out of position by the violent rolling motions during the PIO.

I continued to pull the nose up to decrease my rate of sink until I came level about 50 feet above the lake bed. The vehicle was flying great at this time. It responded promptly and precisely to all my control inputs. I lined up next to the left-hand runway line and then waited until my speed decreased to 240 knots. At 240 knots, I dumped the landing gear out and instinctively responded to the nose-down pitching motion resulting from the landing gear deployment.

The main landing gear doors were relatively large and constituted a significant percentage of the lower fuselage surface. When those doors opened up and the landing gear deployed, there was a substantial increase in drag centered below the vertical center of gravity. This created the nose-down pitch motion and also increased the overall rate of deceleration of the vehicle. I intentionally delayed the deployment of the landing gear to maximize the time available from flare initiation to touchdown. A pilot needed 10 to 15 seconds following the flare to adjust his height and rate of descent prior to touchdown. Delaying landing gear deployment maximized the time available.

We specified a quick-reaction landing gear to enable us to delay gear deployment until the last few seconds before touchdown. Northrop designed a landing gear that would deploy and lock within 1.5 seconds. That's fast. It's also ideal for an unpowered airplane. You can't wait around for the gear to deploy and lock while you are losing airspeed at a rate of 7 knots per second. From the time I deployed the landing gear until I reached an ideal touchdown speed of 200 knots, the interval was less than 8 seconds. As I approached touchdown, the transcript reads:

CHASE 2: One thousand two hundred. Beauty, Milt, you're down. Beauty, Milt.

NASA 1: Beautiful, Milt.

THOMPSON: Rog. No rocket. [I did not use the landing rocket.]

B-52: At least it goes a little farther than the X-15, doesn't it? [The M2-F2 rolled out farther than the X-15 slid out.]

THOMPSON: Yes, it do. I'm trying to get to the Rock-a-Bye. [The Rock-a-Bye was a saloon in Rosamond.]

The nose had swerved slightly as the nosewheel touched down, but the vehicle rolled out reasonably straight. I tried some braking, but the brakes weren't too effective. I decided to let it roll to a stop. When the vehicle finally stopped, I was quickly surrounded by the recovery team vehicles and the rescue helicopter.

M2-F2 POST FLIGHT

Paul Bikle was concerned about the first flight. In fact, he had told Ralph Jackson, the Dryden public affairs officer, that if the M2-F2 crashed, the two of them would walk out the front door and keep on walking. Bikle was putting his career on the line with the first M2-F2 flight. He had advocated the lifting-body program over the objections of his chief of research, Tommy Toll. Toll was never convinced that we should take the risk. He had opposed the construction of the M2-F1, and when he was overruled by Bikle, he recommended that we limit the flight program to ground tows to minimize the risk to the pilot. Toll also opposed the construction of the heavyweight lifting bodies. When he was again overruled, he resigned and went back to Langley Research Center, his original research center. There were other high-level NASA managers who questioned the need for a flight demonstration.

Bikle had some support also. Milton Ames, E. O. Pearson, and Fred DeMerritte were firm supporters. They were the headquarters officials who funded development and construction of the vehicles. Dr. Dryden, the deputy administrator of NASA, had been neutral. He did not want to have an accident and kill a pilot. Bikle didn't want to kill a pilot either, but he believed that the lifting-body concept deserved a real honest appraisal. There was no better way to evaluate the concept than to fly a manned research vehicle. By putting a man in the vehicle, he also demonstrated the concept. If it was a failure, Bikle would take his lumps. He would keep on walking.

Bikle was one of the first to shake my hand out on the lake bed after the flight. There was a real crowd of well-wishers on the lake bed after the landing. The crew finally deactivated the vehicle and jettisoned the remaining peroxide about 45 minutes after landing. We returned to Dryden and conducted a postflight debriefing about an hour later.

The briefing was highlighted by the ground tracking film of the flight. The PIO on final approach was vividly evident. It was a wild ride. The chase aircraft pulled away from the M2-F2 as soon as it began to oscillate and were completely out of the picture before recovery. An over-the-shoulder camera revealed that the vehicle almost went over on its back on the last roll and that my helmet was slammed against the canopy by the high roll rate. I jokingly told the project team members at the debriefing that slamming my helmet against the canopy got my attention and I finally took some action to stop the oscillation. The Northrop flight control engineer couldn't believe that I let go of the stick during that violent oscillation in an attempt to stop it. In reality, I didn't remember my helmet hitting the canopy. I was too busy.

In reviewing the radio communications, we found no communications from the control room during the PIO. They offered no advice, nor did the chase. I was on my own even though everyone was aware of the problem. They may have been able to help, but they were shocked into speechlessness. Thank God, I had planned for such a possibility.

The day before the first flight, Paul Bikle said that the postflight party would begin immediately after the flight. I landed at 11 minutes after 7 in the morning, so it would have been a record-breaking party. I would not have been surprised if the party had started before 8 o'clock in the morning. Bikle usually meant what he said, but alas, the party was delayed. When he greeted me on the lake bed after landing, he had an Air Force visitor with him who was also interested in lifting bodies. I believe it was his visitor who influenced him to delay the celebration until later in the day, although a number of program participants did leave shortly after the landing to begin the party. The party began in earnest in the early afternoon, and it was a memorable one.

Postflight parties are a tradition at Dryden. We were a team that worked hard and played hard. Parties were a way to help vent the pressure that had built up preparing for the flight. We had spent more than a year preparing for this one. Everyone had put in long hours doing the preflight tests, and the ground crew had worked most of the night before launch. Then it was all over in 4 minutes. A high was built up that could not just be shut off. Another purpose of the parties was to build teamwork. There were no unimportant people; the test pilot was the most visible person, but behind him was a large team of people who made it possible but never saw the limelight. Paul Bikle understood. After a flight, he would say, "After you clean up the plane, you can go home."

The first M2-F2 flight received a lot of press attention. The *Chicago Tribune* headlined it, "Wingless craft gliding like brick makes a safe landing." The *National Observer* called it "A flight without wings," while the *Los Angeles Times* added, "Weird post-orbit test craft successful in its first glide." A photo of the M2-F2 in its steep dive made the cover of *Aviation Week and Space Technol-*

ogy magazine. Bill Dana took one look and came up with the perfect caption: "Look Ma, no wings."

REFLECTIONS ON THE M2-F2 PIO

Reflecting back on the PIO, I realized that I was responsible for it in more than one way. During the flight I had moved the interconnect ratio changer in the wrong direction and precipitated an almost catastrophic PIO. Much earlier, however, I had recommended a change in the original configuration of the M2-F2 that caused the vehicle to be PIO prone. The original M-2 design had a set of elevons on the outboard tip fins that were supposed to be the primary roll control surfaces. We duplicated those elevons on the wooden M2-F1 vehicle, and it had no significant PIO tendency. Roll control in the M2-F1 was quite good, once we rerigged the control system.

The elevons were, however, projecting out into the free stream flow and would be subjected to intense heating during an entry from orbit. Most reentry configurations were designed to protect the control surfaces from direct aerodynamic heating to prevent them from burning off or seizing up as a result of bearing failures. The outboard elevons would be subjected to severe gap heating between the inboard edge of the elevon and the outboard surface of the tip fin. There was no easy solution to this problem. We did not have any good high-temperature seals available to close that gap on a movable surface. The only way to solve the problem was to eliminate the elevons and rely on other surfaces for primary roll control. That is what we finally decided to do, since I was a purist and wanted the M2-F2 to represent a real lifting-entry configuration.

Eliminating the elevons wasn't that straightforward. Those elevons provided not only a roll control surface but also some lift. If we eliminated them, we would lose some increment of the lift-to-drag ratio. We were already somewhat marginal on the lift-to-drag ratio, with a maximum value of 3 to 1, so we could not afford to decrease it further. To regain the increment of lift-to-drag, a decision was made to extend the fuselage another 2 feet to the rear. This did two positive things: it increased the overall lifting surface area of the body, and it reduced the depth and therefore area of the base of the vehicle. Those two configuration changes compensated for the loss of the elevons and maintained an overall lift-to-drag of more than 3 to 1.

When we eliminated those surfaces on the M2-F2, we had to use the inboard upper flaps as a roll control surface. The inboard upper flap had two segments, a left and a right segment. The flaps were used in the M2-F1 symmetrically as a pitch control flap only. In the M2-F2, we had to use them as a combined pitch and roll control flap, symmetrically for pitch and asymmetrically for roll. The problem was that these flaps did not produce a pure rolling moment when they deflected asymmetrically. The flap that was deflected upward produced a high-pressure region above it. This high pressure acted on the inboard side of the vertical fin adjacent to it to produce a yawing moment. This yawing moment produced a rolling moment due to the dihedral effect that was as strong or stronger than the rolling moment produced by the inboard flap, and it was an opposite rolling moment. The net result of an asymmetric upper flap deflection was little or no roll response or, in the worst case, a reverse response.

In an attempt to solve that problem, we used an interconnect between the rudder and aileron to deflect the rudder proportional to the upper flap deflection to counter the yawing moment produced by that flap. It was a poor way to solve the problem, but it worked. It did, however, make the M2-F2 PIO prone.

There are numerous reasons for a PIO tendency in an aircraft. Generally, however, the cause is a deficiency in either its aerodynamic characteristics or its control-system mechanization. An aircraft doesn't have to be deficient in both aerodynamics and controls. It may be a good airplane aerodynamically but have a poorly designed flight control system.

The M2-F2 design required an unusual set of control surfaces. To compensate, we tried to develop a control system to work around it. On the first flight we demonstrated that the M2-F2 could be flown safely into the predicted PIO region with the proper aileron-rudder interconnection ratio, and even recover from a PIO should one be encountered. Despite this significant success, we would ultimately learn that eliminating the M2-F2 PIO problems required modifying the aircraft's aerodynamic configuration.

The PIO on final had been triggered by my turning the interconnection ratio changer in the wrong direction. That error had been caused by the difference between the ratio changer in the M2-F2 and the simulator—a bad mistake on the part of the simulator crew and myself for not insisting on the proper device. An aircraft's controls should always be reproduced accurately in the simulator cockpit. Many serious problems have been encountered when the two sets of controls didn't match. The pilot spends many hours in the simulator preparing for flight, especially the first flight. He learns to control the vehicle instinctively. He doesn't have to look at a control to know what it is or which way he should move it to get the desired result. He adjusts various controls by feel rather than sight. If the simulator does not duplicate the airplane, the pilot has to stop and think about what control lever he is touching and which direction he must move it. That could mean a deadly delay or a deadly error, as it almost did in this case.

The successful first flight of the M2-F2 was a major milestone in the validation of the lifting-body concept. It demonstrated that these oddball shapes could be maneuvered to a precise location and landed successfully by a pilot without the need for landing engines. It made lifting bodies much more competitive with ballistic capsules for manned spacecraft. They were somewhat more complex, with their aerodynamic stabilizing and control surfaces, but they did not require a parachute or parawing for landing. They could be landed horizontally just like any other airplane, even without engines. The first flight of the M2-F2 was the first step in persuading NASA to select a lifting-entry configuration for the space shuttle orbiter rather than a ballistic capsule such as the Big Gemini.

The first flight of the M2-F2 did not answer all of the questions about lifting bodies. There was a lot of work to be done to demonstrate their stability and controllability in the transonic and supersonic flight regimes, but that first flight was a mighty step in the right direction.

FLIGHT M-2-9

The second flight of the M2-F2 was scheduled for July 19, one week after the first flight. This week-long interval allowed time for a thorough check of the vehicle, some minor repairs, and a minor modification to the landing rocket propulsion system. The preparation for the next flight involved a thorough checkout of each system in the vehicle and a servicing of these systems as necessary.

My preparation for the second flight involved a lot of simulation to compare the vehicle's simulated handling-quality characteristics with those observed on the first flight. If there were some significant differences, the simulator was modified to make it fly like the real airplane. Some of this comparison was done on the basis of my qualitative evaluation. Some was based on aerodynamic characteristics computed from the first-flight data. It was vital that the simulator be improved as good flight data became available to assure that it would accurately predict the vehicle's characteristics at the next test point. We did not want to be flying off a cliff. A more realistic simulator could possibly allow us to see the cliff coming up.

Once the simulator was updated, the flight planner and I developed the next flight plan using the simulator. Then the practice sessions began to ensure that I was intimately familiar with every aspect of the planned flight. During these practice sessions, the flight planner would vary some of the vehicle's predicted aerodynamic characteristics to familiarize me with the resulting effects on handling qualities. This was a mechanism to alert me to any discrepancies between the simulator and the vehicle at the next data point. If there were major discrepancies, it might be necessary to back off or approach a new test point more conservatively.

Another part of the preparations for the next flight involved practicing approaches and landings in the F-104 aircraft. This included simulated landings on the intended runway and any other runway that may be used in case of emergency. It required almost a week to accomplish all the training, vehicle inspection, and vehicle preparation.

The M2-F2 was mated to the B-52 on July 16. Postmate checks were performed on July 18. The vehicle was serviced with propellants and source gases early the next morning.

The proposed flight plan was to be a rather simple one to evaluate lateral stability and control, to conduct further systems checkout, to measure vehicle performance, and to verify longitudinal trim.

The ground track would involve only two legs, a crosswind leg and a final approach leg. They were relatively long legs to allow me to conduct more data maneuvers without having to maneuver excessively for landing. The launch point was to be approximately 6 miles north of Boron on a westerly heading roughly parallel to the north shore of the lake bed. I could maintain this westerly heading until I was almost lined up with Runway 18-36 on the north lake bed. At that point I would make a left turn to line up with the runway and then fly straight on it to a landing on Runway 18.

The maneuvers to be performed included aileron step inputs with varying aileron-rudder interconnect ratios to verify a predicted roll reversal response at low interconnect ratios. These maneuvers were to be performed at low and moderate angles of attack. Other maneuvers included rudder and pitch stick pulses to obtain data for determination of the vehicle's aerodynamic characteristics such as control effectiveness and static and dynamic stability. These maneuvers were also to be performed at two different angles of attack.

The flight proceeded as planned. Everything went smoothly during pilot entry and cockpit checks. We taxied out on time. I did comment on the smoother ride during taxi. On the previous flight, it felt like the B-52 had square tires; on this flight, they appeared to be nearly round. Flat spots developed on the B-52 tires after sitting for long periods with a load on the airplane. To make matters worse, the first few thousand feet of the runway, where the takeoff roll occurred, were rough. It was a jolting ride out on the wing at times.

At 15 minutes to launch, I started through the prelaunch checklist to begin activating various M2-F2 systems and then checking for proper operation of the systems. The checklist was rather lengthy, but not as lengthy as the one for the X-15. There were fewer and simpler systems in the M2-F2. I activated the hydraulic systems, which provided pressure to the flight control systems, and then began checking the control surface positions, stability augmentation system configuration, cockpit circuit breakers to make sure they were all in, and then began calling out the cockpit gauge readings, such as the altimeter, angle of attack, and angle of sideslip, to cross-check these with the information that was being displayed in the control room.

Following that, we began a series of control movements to verify that we were getting full deflection of the control surfaces under airloads. We then checked for proper SAS operation by having the B-52 pilot pulse the controls to create specific motions of the M2-F2 on the wing pylon. Some of those pulses were strong enough to rattle my teeth. We spent a lot of time verifying the correct position of the control surfaces prior to launch. We were concerned about this, as the M2-F2's motion after launch was highly dependent on control surface trim. The M2-F2 was much more susceptible to violent motions during launch than the X-15, since it had little inherent roll damping. We didn't want to do a series of barrel rolls after launch.

At 7 minutes to launch, I began calling out system gauge readings such as helium gas pressure, governor balance pressure, cabin air pressure, and peroxide tank pressure. At the same time, the B-52 began its turn back toward the launch point. I checked the landing rocket at 5 minutes to launch, switched to the M2-F2 radio, and made a radio check with NASA 1.

At 4 minutes, I went to M2-F2 defog source and emergency position to get maximum defogging action. I really didn't want the canopy and nose window to fog or ice up during the descent—that could be catastrophic. NASA 1 called to ask about the forward window defog switch, and I confirmed that I had both canopy and forward window defog switches to the emergency position. I then switched my oxygen supply from the B-52 to my own oxygen tank in the

M2-F2 and called out the oxygen pressure in the tank. I also switched my cabin air source from the B-52 to my own source in the M2-F2.

At 3 minutes to launch, the B-52 was still in its turn but approaching the roll-out heading of 276 degrees. At this time I switched my cabin heater supply from the B-52 to the M2-F2. All of this switching was in preparation for the launch. We wanted to be using our own internal supplies in the M2-F2 just before launch so the supplies would not be interrupted during launch. It would really get my attention if I were switched to B-52 oxygen at launch because the supply would stop. I wouldn't get any breathing oxygen until I switched to the M2-F2 oxygen. We normally relied on the B-52 supplies of various consumables like oxygen, cabin air, and electrical power as long as possible since the B-52 could carry much more than we could in the M2-F2. The M2-F2 consumables were quite limited in duration, normally just enough to accomplish the free-flight mission.

Between 3 and 2 minutes to launch, there were a number of B-52 heading changes to line up on the desired launch point. When NASA 1 called 2 minutes to launch, I turned the cockpit camera on and then checked my stability augmentation system status to make sure it was functioning properly. After a couple more calls to adjust the B-52 heading, NASA 1 called 1 minute, "Now." I confirmed the 1-minute call and then uncaged my gyros and checked my cockpit camera switch. I was now ready for launch.

At 5 seconds to launch I counted down, "Five, four, three, two, one, release." NASA 1 called immediately to have me check my dampers, or stability augmentation. I confirmed that they were still on and functioning properly. I then began setting up for the test maneuvers. The first test maneuvers were aileron inputs at three different interconnect ratios to determine roll response characteristics. The three interconnect ratios were 0.2, 0.5, and 0.8. The simulator predicted that the roll response would be overly sensitive at 0.8, about right at 0.5, and too sluggish at 0.2. The actual vehicle roll responses were about as predicted. I then did some pitch, roll, and yaw pulses with the dampers on and off. These showed little or no differences in damping with them on or off. The vehicle was flying well.

At 30,000 feet, I began my left turn onto final for Runway 18. As I rolled out on final, I pushed over to decrease the angle of attack and increase airspeed. On reaching 215 knots or about 2 degrees angle of attack, I again made some abrupt roll inputs at 0.8, 0.5, and 0.2 interconnect ratios. At 0.8, the roll response was overly sensitive. At 0.5, it was good. At 0.2, the vehicle rolled in the wrong direction. This was predicted by the simulator, but it was an uncomfortable response when it occurred in flight. I was anticipating that the vehicle would roll the wrong way, but when it did, my immediate reaction was to put in more roll to stop it from rolling the wrong way. This additional roll input just aggravated the problem, since the vehicle rolled the wrong way even faster. It took a tremendous amount of willpower to stop increasing the roll input. I couldn't convince myself that applying a roll input in the wrong direction would stop the roll and bring the wings back to level.

The X-2 was lost as a result of this phenomenon. I could now understand why. In the case of the X-2, when the pilot, Captain Apt, began his turn back toward the lake bed after engine burnout, directional stability decreased to zero as the angle of attack increased in the turn. As the directional stability decreased, the adverse yaw produced by the ailerons began yawing the aircraft nose in the wrong direction. This nose yaw created a rolling moment in the wrong direction due to dihedral effect. The pilot increased his aileron deflection in an attempt to roll the aircraft in the right direction, but in doing so, he gravely aggravated the situation, and the aircraft quickly departed controlled flight. The pilot was unable to recover and finally separated the cockpit from the fuselage in an attempt to bail out. The cockpit was designed to be a high-speed escape capsule that would protect the pilot until it slowed down to a speed where he could bail out. The escape capsule worked as designed, the drogue chute came out to stabilize it, but the capsule rolled violently on the end of the drogue chute riser. The pilot was battered by this violent rolling motion and did not attempt to bail out until it was too late. I could now understand why he reacted the way he did when the aircraft began rolling in the wrong direction. I might have done the same under those conditions.

The remainder of my flight went smoothly. I had to make a large S-turn to kill off some energy during final approach, but the flare and landing went smoothly. Just before the landing, the chase noticed that the nose gear was oscillating from side to side but then straightened out just before touchdown. As the main gear touched down, the left strut deflected more than the right, and when the nose gear touched down, the vehicle swerved sharply to the left momentarily and then recovered and rolled out in a gradual left turn that decreased in radius as the speed decreased. I finally stopped after turning more than 90 degrees from the runway heading. That is not a comfortable landing rollout, particularly when landing at such high speeds. The landing gear configuration on the M2-F2 was sensitive to uneven main gear strut deflections.

The vehicle would turn into the low strut, and there was little the pilot could do to counteract it. We had no nosewheel steering, and the brakes were ineffective. If the aircraft landed with a low main gear strut, the pilot was just along for the ride.

FLIGHT M-3-10

The purpose of the third flight, scheduled for August 12, was to increase the launch speed from 0.6 Mach number to 0.65 Mach number and to evaluate the M2-F2 motions during the launch. We wanted to build up to a launch Mach number of 0.8, if possible, to maximize our energy prior to launch and enable the M2-F2 to achieve a maximum Mach number after launch. Our first launch was made at a low Mach number to minimize the M2-F2 motions after release. Launch transients are generally greater with increasing speeds.

The other purposes of the flight were to determine the minimum acceptable levels of damping required of the stability augmentation system, and then to obtain stability and control data to compare with wind tunnel predictions. We wanted this information to update our simulator, so it would predict the M2-F2's flight characteristics accurately. Updating the simulator was essential when testing a new aerodynamic configuration, particularly an unconventional one like the lifting body, to allow us to improve our prediction of flying characteristics at the next level of speed, angle of attack, or Mach number.

The ground track for this flight was to be similar to the first flight's track: a short straight leg after launch and then a left 90-degree turn, followed by another straight leg, and then finally a 90-degree turn onto final. The launch point was again just east of the rocket site. We abandoned the L-shaped ground track used on the second flight, since it was not easy to judge energy during the landing approach. I had been too conservative in managing my energy during the landing approach, and as a result, I had had to make a rather wide S-turn to kill off my excess energy prior to landing.

Prior to launch, the systems checks went well, with only minor anomalies. I did have some radio interference from George Air Force Base that blocked out a few radio transmissions from NASA 1, but the radio was good after launch.

The launch this time felt more like an X-15 launch, with an abrupt deceleration after launch, like I had hit bottom. The roll-off this time was faster and larger. I rolled off 40 to 50 degrees in bank angle before the vehicle stopped. It came back to a wings-level attitude readily, however. I then began setting up for the control pulses, but the vehicle wouldn't stay trimmed. I finally got it trimmed and then made some rudder pulses. The vehicle response wasn't very predictable. I had the impression that there was some slop in the rudder linkage that gave a nonlinear response to a rudder pulse. I made the first turn with the roll and yaw dampers off, and the vehicle flew quite nicely.

The final approach and landing were comfortable. I was S-turning to dissipate some energy when I passed through the 1,200-foot flare initiation altitude, but I felt confident that I could delay the flare and still come level at the desired height above the ground. I now had enough experience in the M2-F2 to fly it a little more like the X-15 in the landing maneuver, and I was willing to flare at a lower altitude using a higher g level. I came level about 20 feet above the lake bed, waiting until the airspeed bled off to 230 knots, and then deployed the landing gear. I corrected for the pitch-down tendency and finally touched down at 200 knots.

There was no swerve as the nose gear touched down, and the vehicle rolled out reasonably straight until it slowed down, and then it began swerving to the left. I crossed all three runways before I finally got the vehicle stopped. I used a lot of braking as I crossed the last runway because I knew there were some big potholes to the east of the three marked runways on the lake bed.

In the postflight debriefing, the chase pilot indicated that the nosewheel was oriented 90 degrees to my flight path just before touchdown, but it straightened out just before nosewheel touchdown. Apparently we still hadn't solved the nose gear misalignment problem. The tendency for the M2-F2 to swerve left during rollout was attributed to a crosswind of 10 to 15 knots on the lake bed. This vehicle, like the X-15, would drift downwind as it decelerated, and there was little anyone could do to stop it.

FLIGHT M-4-11

Flight 4 (M-4-11) was made on August 24, 1966, and was intended to expand the launch speed to Mach 0.7 and to obtain more stability and control data. We also planned to obtain some airspeed data to calibrate the system. The flight path was the same as for flight 3, with a planned launch just east of the rocket site.

On this flight, however, we encountered numerous radio problems that caused us to abort the launch and make a 15-minute holding pattern for a second launch attempt. The 15-minute hold had a noticeable effect on hydraulic system operation. The hydraulic system was sufficiently cold-soaked during the holding pattern to prevent the normal buildup of pressure once the pumps were turned on. Other systems showed some subtle effects of the additional cold-soak. The outside air temperature at the launch altitude of 45,000 feet was roughly minus 65 degrees F. That is cold. My feet and legs really got cold before this launch, since we had poor circulation of air in that area of the cockpit.

The flight went smoothly after launch until the final turn into the runway. I was higher than planned and again I had to S-turn vigorously to kill off energy and land at the designated location on the runway. After landing, the vehicle rolled out perfectly straight. There was no swerving at nose-gear touchdown or during the rollout. We had apparently solved our steering problems.

FLIGHT M-5-12

Flight 5 was planned to demonstrate the M2-F2's maneuverability in the landing pattern. The optimum landing pattern for an unpowered aircraft is a 360-degree pattern. This was the normal pattern for the majority of the X-15 flights. In a 360-degree approach pattern, the vehicle arrives over the desired touchdown point at a predetermined high key altitude on the same heading as the landing runway. After passing over the planned touchdown point, the pilot initiates a right or left turn to arrive at a position abeam of the touchdown point at a predetermined low key position heading in the opposite direction to the runway. The turn is stopped for a short period at the low key position to fly downwind a couple of miles before again commencing the turn. This downwind position is desirable to provide a straight-in final approach of a mile or two prior to the final flare for landing.

During the final approach, the pilot is adjusting his airspeed and aim point to ensure a precise touchdown point at the proper airspeed. This type of pattern was used quite successfully in the X-15 to achieve touchdown accuracies of less than 1,000 feet from the planned touchdown location. The approach pattern in the M2-F2 was similar to the X-15 pattern except that it involved a steeper descent rate and a tighter circle or turning radius. This was partly because of the lower gliding performance of the M2-F2. The M2-F2 had a maximum lift-to-drag ratio of 3; the ratio for the X-15 was 4. Final approach flight path angles were nominally minus 27 degrees for the M2-F2, versus minus 18 degrees for the X-15. The M2-F2 was coming down like a dive-bomber. In fact, after watching a movie of an M2-F2 approach and landing, one of the German engineers working at the Marshall Spaceflight Center compared it to the German Stuka dive-bomber—"Yaaah, joost like der Stookas."

We made the fifth flight on September 2, 1966, and it went smoothly. The launch was mild, with a maximum roll-off of 25 degrees. After leveling the wings, I turned the roll and yaw stability augmentation off and then rolled into the turn toward the low key position. The M2-F2 flew quite well with the SAS off. I tried a couple of different control techniques to determine which would be the most desirable in case the SAS really did fail. After passing low key, I reengaged the roll and yaw SAS and disengaged the pitch SAS. Again, the M2-F2 flew quite satisfactorily without pitch augmentation. Pitch damping would still be required during launch and during landing gear extension to prevent the M2-F2 from departing into uncontrolled flight.

The final approach on this flight was similar to the approaches I had made on the previous flights. I was again high on energy turning onto final and had to S-turn to dissipate the excess energy. I held the M2-F2 off the ground longer than usual before touchdown to determine a minimal touchdown speed in case of a last-second landing gear deployment problem or a low-energy approach problem. I managed to hold the M2-F2 off the ground until I had decelerated to 155 knots. That provided an additional 6 to 8 seconds of float time to cope with a last-minute problem. Not much time, but it might be enough to prevent an accident.

THE M2-F2 GLIDE FLIGHTS: A SUMMING UP

These first five glide flights had accomplished quite a bit even though they were only short flights, averaging approximately 4 minutes each. They demonstrated that we could fairly accurately predict the flying qualities of these oddball lifting-entry configurations at subsonic speeds. The M2-F2 exhibited poor handling qualities at low angles of attack because of its PIO tendency, but the wind tunnel data and simulation had predicted that. These results would imply that we could successfully predict the flying qualities of other lifting-entry shapes with equally strange configurations.

The five flights also demonstrated that these oddball shapes could be successfully maneuvered and landed accurately at a preselected location without landing engines—an important result. Landing engines would add complications to a spacecraft design and would also impose a large weight penalty that would significantly reduce the allowable payload or require a much larger vehicle to carry the same payload as a spacecraft without landing engines.

Finally, the five flights had demonstrated the feasibility of a lifting-entry spacecraft and had made a lifting-entry configuration competitive with capsule designs for future spacecraft. We still had to demonstrate that these lifting bodies had adequate transonic and supersonic flying characteristics, but we had taken a giant step forward with these five subsonic flights.

FROM COCKPIT TO COUCH POTATO

I had told Paul Bikle that I planned to quit flying and move back into the research organization after I had flown the M2-F2. We hadn't agreed on exactly when I would quit, but he was anxious to have me move back into research engineering. Bikle told me years later that he was always concerned when a pilot indicated that he was going to quit flying. He had observed that a number of them had been killed in accidents shortly after they had made the decision to quit but before they had actually quit. Bikle indicated that he almost grounded me after I told him that I planned to quit; he wanted to save me from myself. He said that he allowed me to fly the M2-F2 only because he believed that I would not do anything foolish or hazardous. Coming from Bikle, that was a big compliment.

I did have second thoughts about giving up test flying. I really enjoyed the challenge and the excitement. There was always something new and interesting happening each day, but I could not see any new exploration programs coming along soon that would really turn me on. There was no planned follow-on to the X-15 or lifting bodies. I wasn't too interested in boring holes in the sky, so it was time to quit.

My decision took a lot of tension out of the family. I still had the Jag when I was with the X-15 program. Eric, who was then 12 or 13 years old, would play in the car, pretending to drive. One time he found an envelope stuck down between the seats. Inside was a piece of paper designating the executor of my will. I had written it several days before an X-15 flight, but it had the date of the flight. After retiring, there was also not as much travel. They could count on my being home at 5 or 6 P.M. most nights. I was there for night or weekend school functions. After arriving home, I would put on a T-shirt and jeans, then mow the lawn or work on the car. I also spent many hours helping Peter with his math homework.

I spent several months vacillating on my move back to engineering. Prior to initiation of the M2-F2 powered flight program to achieve higher speeds, we had to check out Bruce Peterson, Don Sorlie, and Jerry Gentry to carry on the flight program. I continued to act as chief lifting-body pilot until the new pilots were checked out, but I knew I had to decide soon. Bikle would stop by occasionally and attempt to nudge me into a decision, but I still couldn't cut the umbilical cord. Bikle proceeded to prepare a new office for me in research and finally shamed me into establishing a date for my move. I moved out of the pilots' office three months after my last M2-F2 flight. It was a traumatic decision. I'm happy that I made that decision, but it was the toughest decision I ever had to make.

Bikle offered to let me continue to fly our support aircraft, but I decided I had better make a clean break and quit flying completely. Otherwise, I might be tempted to reverse my decision. Bikle solidified my decision one day by informing me that I had to give up my key to the pilots' squash court. That was a real blow to my ego. The squash court was a status symbol for the pilots. It was originally constructed to provide the pilots with a means to stay in shape physically. It was used continuously by all the pilots and did serve its intended purpose. The pilots were in good shape physically. Now I would become a couch potato.

8
HARD TIMES

The next step in the M2-F2 program was to check out the new pilots and then prepare to begin powered flights. As with the X-15, the lifting-body program was run as a joint program of NASA and the Air Force. The NASA pilot was Bruce Peterson, while flying for the Air Force was Jerry Gentry. Because of Gentry's two slow rolls while flying the wooden M2-F1, his checkout was put on hold until Don Sorlie had flown the M2-F2 and given his approval.

Peterson made his first glide flight on September 16, two weeks after my final flight. Don Sorlie made his first lifting-body flight on September 20. Peterson made his second flight in the vehicle after a quick, two-day turnaround, on September 22. Sorlie made the next two flights and cleared Gentry to join the program. Gentry made the next four flights in October and November 1966. After his November 21 flight, the fourteenth for the M2-F2, the vehicle was grounded for installation of the XLR-11 rocket engine.

In the course of the checkout flights, the M2-F2's tendency toward pilot-induced oscillation had reappeared twice. On one of Don Sorlie's flights, a PIO had occurred during a research maneuver at altitude. The specific maneuver called for did encroach on the PIO flight region. I criticized the flight planners for selecting that specific maneuver at those flight conditions, but they felt that Sorlie could cope with any problem that might arise. He was able to, but not before he lost control in a mild PIO.

The second PIO occurred during Jerry Gentry's checkout flight. We had scared Gentry to death about the PIO and the need to stay away from the critical flight region. As he rolled out on final approach, he felt the slight nibbling in the control stick that warned of an impending PIO. He immediately pulled up, which increased the angle of attack and eliminated the PIO. Gentry then continued his approach at a lower airspeed or a higher angle of attack. He had no problem completing the final flare and touchdown.

We did have some excess airspeed built into our specific approach and landing airspeeds. We had put special effort into Gentry's checkout flight. Gentry had demonstrated an ability to react rapidly to any aircraft motion. His nominal reaction time compared well with the stability augmentation system response time. We referred to him as a human rate damper.

A fast reaction, however, isn't always the best response. It can incite a PIO or other undesirable aircraft motion. Most pilots tend to filter out these extremely high-rate responses and allow the SAS to do the job it was intended to do.

These PIOs were the first indication of the hard times that would soon beset the lifting-body program.

THE HL-10 FIRST FLIGHT: THE START OF HARD TIMES

While the M2-F2 was undergoing glide flights, the HL-10 was being prepared. The configuration had evolved at the Langley Research Center separate from the Ames lifting-body studies. While the Ames lifting bodies used a half-cone shape (rounded on the bottom and flat on top), the Langley studies under Gene Love and his colleagues used a more complex shape. The underside was flat, but the nose and tail angled up. The top surface was straight, while the sides curved into the flat underside. Two tip fins and a center fin provided stability, while flight control used split flaps like on the M2-F2. One difference in the flap arrangement was that they had two different positions, raised for transonic flight and flat for subsonic.

The HL (for "horizontal landing") shape had the advantage of being stable in the pitch, roll, and yaw axes at angles of attack up to about 52 degrees and retaining a higher lift-to-drag ratio at low angles of attack. Between 1962 and 1964, the basic shape underwent a series of wind tunnel studies with different fin configurations. The tenth shape (thus "HL-10") had better handling qualities and a higher lift-to-drag ratio than the M2-F2. It was also the best looking of the original lifting-body shapes.

The HL-10 was delivered on January 18, 1966, six months after the M2-F2. The new research aircraft stretched our resources to the limit. The X-15 program was still under way with three airplanes, and it was occupying a large percentage of the Dryden personnel. As a result, authority for the lifting-body program was split between NASA and the Air Force. The Air Force would be responsible for M2-F2 flight planning and simulation. Dryden would maintain responsibility for aerodynamic data, stability and control, and all instrumentation and maintenance. That freed the Dryden simulation facilities and engineers to concentrate on preparing the HL-10.

The Air Force M2-F2 and NASA HL-10 teams had different personalities. The program manager of the Air Force team was Bob Hoey, while Johnny Armstrong was the project engineer. Hoey had been at Edwards about twelve years, Armstrong ten years. Both had been closely involved in the X-15 program and had a large credibility base with Paul Bikle and within Dryden. They were considered the experts.

The HL-10 team was, in contrast, new and unproven. Garry Layton was the project engineer, Wen Painter and Berwin Kock were the HL-10 system engineers, Jon Pyle handled vehicle performance, and Bob Kempel was responsible for stability, control, and handling qualities. Each had only three to six years of experience. This gave rise to doubts, which became worse when the HL-10 simulator became operational.

According to the simulator, the HL-10 was more stable and would have much better handling and lift-to-drag ratio than the M2-F2. The pilots who flew the HL-10 simulator were suspicious and skeptical of the results, saying it was "too good." This opinion was shared by John McTigue (the lifting-body program manager), Paul Bikle, and the Air Force M2-F2 team. Managers would pass the HL-10 team members in the hallways and shake their heads. A common comment was "It can't be that good."

The HL-10 was sent to Ames for wind tunnel testing, then spent months undergoing ground tests similar to those done to the M2-F2. Two captive flights were made under the wing of the B-52. Shortly before Christmas, the HL-10 team indicated the vehicle was ready to make its first glide flight. Flight H-1-3 was scheduled for December 21, 1966, but a problem with a tip fin flap electrical circuit caused it to be aborted before takeoff.

The next day, December 22, the repairs were complete and Bruce Peterson climbed into the HL-10 and began the preflight checklist. When it was completed, the B-52 taxied out to the main runway and took off. The launch point was about 3 miles east of the east shoreline of the lake bed, almost directly above the rocket site. The HL-10 would be released abeam of Runway 18 heading north, then make two left turns, the same flight path we had used for the

M2-F2: a downwind leg, a base leg, a turn to final, and a final approach to Runway 18.

The HL-10 was launched at 10:38 A.M. from the B-52 while flying at 45,000 feet at a speed of 170 knots. The launch and aircraft trim were much as expected from the simulator. The pilot, however, became aware of a high-frequency buffet in pitch and some in roll. The whole vehicle was moving and shaking. As the vehicle descended and picked up speed, the buffet became worse.

As the first turn was made, the pilot noticed that the stick was very sensitive in pitch. A small stick movement resulted in far too much change in pitch. As the high-frequency buffet increased with the vehicle's speed, it became obvious that the stick was also too sensitive in the roll axis. The pilot made several changes in the pitch gain to try to control the sensitivity, but the problem kept getting worse. Nothing seemed to help.

One of the planned maneuvers was a practice flare at altitude. As Peterson slowed the vehicle and pulled the nose up, the vehicle stopped responding to roll control. Peterson found himself with the stick full left and aft with no response from the aircraft. He pushed over early, and the control problems eased. This also raised the preflare airspeed from the planned 300 knots to around 340 knots. The pilot could evaluate the vehicle's handling while he was still high enough to bail out if it should prove uncontrollable.

The control problems were most severe during the final third of the flight. The oversensitive pitch reappeared, and Peterson reduced the gain again. The setting had been reduced from 0.6 and was now at 0.2, the next to last setting. In the control room, the engineers, now aware of the control problems, feared that a PIO might occur when the landing gear was extended. Peterson realized the only way he could make a landing was to keep the angle of attack low to keep the speed up, and touch down before the roll, and possibly pitch, control was lost. The flare was made at about 320 knots and completed only 30 feet above the lake bed. The touchdown was made at about 280 knots, 30 knots faster than planned. Total flight time was a long 3 minutes and 9 seconds.

Peterson was concerned about both the buffeting and the oversensitive stick. He judged the deficiency warranted improvement and was not satisfactory until improved. The flight was also disappointing to the HL-10 team. The skepticism expressed by the experts was justified, as the HL-10's flying qualities were quite poor compared with the simulator. The HL-10 team felt as if the Air Force M2-F2 personnel were looking down their noses at them.

The arrangement of the data transmitted from the vehicle in flight to the engineers on the ground was also haphazard. This made it difficult to see what was happening in either real time or during the postflight analysis. As Peterson first encountered the control problems, the engineers on the ground could not see that anything unusual was happening. Certain physical relationships existed between different sets of data. When organized together and viewed as sets, what was happening to the vehicle became clearer. Just as we had underestimated the PIO problem in the M2-F2, the poor organization of the data caused the HL-10 team to overlook subtle problems with that vehicle.

POSTFLIGHT ANALYSIS

The Christmas holidays were only days away, and the HL-10 team took several days off before beginning the postflight analysis. When they returned, it initially appeared that the problems in the first flight were easily solved. The buffeting in the pitch and, to a lesser degree, the roll axis was caused by a higher-than-predicted elevon effectiveness, which caused the feedback of a 2.75-Hz limit cycle oscillation into the SAS. This was cured by adding a filter to the SAS electronics. The stick sensitivity problem was traced to the gearing of the stick. This was reduced by half, which resulted in less movement of the control surfaces for a given stick movement.

The stick sensitivity problem was hard to anticipate because of the limited simulator technology of the early and mid-sixties. We used a fixed-base simulator, which gave little in the way of visual or motion clues. As I had learned in the M2-F1 and M2-F2 simulators, this made it difficult to determine exact flight characteristics.

Peterson and the HL-10 team were satisfied that they had solved the problems and were ready for a second flight. Paul Bikle gave his approval. The exception was Wen Painter. He was convinced something was being missed. As a result, the second flight was put on hold pending further analyses. Bob Kempel compared the flight data with the computer's mathematical model of the HL-10. He took twelve flight maneuvers and compared them with the computer's predictions. Of the twelve, only seven maneuvers matched, and they were only marginal. In the other five, there was no similarity between the computer results and the flight data.

The flight data were rearranged in a logical manner to show the relationships between the sets. When this was done, it became apparent that during two intervals the pilot had commanded significant amounts of aileron, but the vehicle did not respond until the angle of attack was reduced. As the angle of attack was reduced below the values of 11 and 13 degrees, the ailerons suddenly became very effective, producing significant amounts of roll (30 to 45 degrees per second). This particular problem had not been apparent to the pilot, and he had made little or no comment about it during or after the flight. When Kempel tried to match the computer results with the flight data, he discovered that the initial part did not match, but when the angle of attack was reduced and the ailerons became effective, the mathematical model began to match the flight data.

The HL-10 team began to suspect that the cause was a massive flow separation over the aft fuselage at high angles of attack. Normally, the air moved over the surface of each tip fin in a smooth, laminar flow. This generates an inward-pointing force, which stabilizes the vehicle, much as the airflow over a wing generates lift. Flow separation is similar to the stalling of a wing, when drag increases and lift decreases. The air is no longer flowing smoothly over the tip fin's surfaces. The flow becomes detached from the fin, and the fins can no longer stabilize the vehicle, nor can the control surfaces act on the airflow to produce a roll moment. When the angle of attack is reduced, the airflow suddenly reattaches and the controls again become effective.

Langley was notified of the team's suspicions and immediately agreed to put a scale model of the HL-10 into a wind tunnel. This was unusual, as wind tunnel schedules are normally made a year or more in advance, but the HL-10 was Langley's baby. Wen Painter, Berwin Kock, Bob Kempel, and Garry Layton went to Langley to present their data and hypothesis. At the meeting were Wayne McKinney, Bill Kemp, Tommy Toll (who had opposed the manned lifting-body program), and Bob Taylor of Langley.

As the HL-10 team members presented their hypothesis of flow separation, Bob Taylor jumped up from the table, angrily threw his mechanical pencil to the floor, and began swearing. Everyone in the room was shocked by the outburst. When Taylor calmed down, he explained, "I knew that this would be a problem." Some flow separation had appeared in the wind tunnel data, and he had a gut feeling that it would be worse in flight. Taylor was upset with himself for not following his instincts and taking corrective measures before the vehicle was built. The HL-10 was grounded, and Langley promised to give the problem immediate attention and propose modifications.

In the meantime, we had bigger problems. The future of the whole lifting-body program was in question.

FLIGHT M-16-24

The M2-F2 spent five months in the hangar for installation and ground tests of the XLR-11 rocket engine. Two glide flights were planned to test the handling of the vehicle with the added weight of the rocket engine and the modifications needed to fit it in the vehicle. This required that the lower flap, used for elevation and pitch control, be bumped out. The upper flaps then had to be raised to compensate. The first powered flight would be made on the seventeenth launch.

Jerry Gentry made the vehicle's first glide flight with the engine installed on May 2, 1967. It was his fifth flight in the vehicle. He was no big fan of the M2-F2's handling qualities, and this flight intensified his dislike. After launch, he found the changes in the flap arrangement caused an increase in drag, which lowered the vehicle's already limited performance. Several times during the descent Gentry thought he was on the verge of losing control of the vehicle. On final approach, he again felt he was not in control. The landing was successful despite the problems. At the postflight debriefing, Gentry said that he did not want to fly the vehicle again until it was fixed. The decision was made to have Peterson fly it and see if he thought it was really that bad.

Flight M-16-24 took place on May 10, 1967, and was to be the last before a powered flight was attempted. Bruce Peterson was the M2-F2 pilot, ground controller was Jerry Gentry, and Chase 2 was John Manke in an F5D. Takeoff and prelaunch preparations were normal. Launch took place at 9:38:21.6 PDT. The flight was normal from launch to the final turn toward the lake bed. The research maneuvers were conducted on the first two legs of the U-shaped flight path. Several times, however, Peterson indicated that he was not satisfied with the control gain settings. At one point he said, "I think my interconnects are riding a little bit too high." Soon after, he said, "I'll raise the gains. I don't like that too well. I'll lower my gains a little bit."

As the vehicle came out of the left bank toward the landing leg, the pilot said, "Boy, there's some glitches." The vehicle was at a low angle of attack and thus very sensitive to lateral control inputs. About the time of the call, Peterson recalled later, the nose moved out to the right, and he thought a wind gust had hit the vehicle. The pilot instinctively used the rudder to control it, even though he knew that the rudder should not be used in low angle-of-attack conditions. The rudder aggravated the oscillations, and the vehicle made two rolls to a bank angle of 140 degrees at a roll rate of 200 degrees per second. Manke in Chase 2 radioed, "Okay, Buddy, when she rights up, get out." Peterson stayed with the vehicle and brought it under control by pulling back on the stick and increasing the angle of attack.

But the loss of control meant the vehicle's heading was considerably more southerly than planned, and there was no time left to make any changes before making the landing flare. As Peterson began the flare, he saw a rescue helicopter directly in front of him. The pilot made three radio calls regarding it. Manke reassured him, "He's all right, you're okay, you're clear, watch your gear, Bruce."

The M2-F2 cleared the helicopter by several thousand feet, but as the flare was completed the vehicle hit the lake bed with its landing gear still retracted. As the underside contacted the lake bed surface, the telemetry antenna was sheared off. In the control room, all the instrument readings went to zero, and several engineers looked up at the television monitor. The vehicle bounced back into the air, and Peterson pulled the gear handle and apparently fired the landing rockets to try to buy time. The vehicle hit the lake bed about 80 feet farther down the landing path. The marks in the lake bed indicated the gear had partially deployed before the vehicle hit. After skidding a short distance, the vehicle turned sideways and began rolling. As it went over, the landing gear had fully extended. The vehicle made six rolling revolutions, which tore off the landing gear, canopy, and right vertical fin. It finally slammed into the lake bed upside down, resting on the left fin and the seat support. The last thing Peterson remembered before blacking out was a high deceleration and the vehicle starting to roll at an increasingly faster rate.

The helicopter was the first to reach the vehicle. Jay King, a NASA crewman aboard the helicopter, jumped out and ran over to the vehicle to help Peterson out. When King crawled under the wreckage, he found Peterson hanging upside down in the straps with severe facial injuries. King thought he might be dead. He turned away and took some deep breaths and then heard Peterson call out, "Help me." King unfastened the harness, and he and Joseph Huxman, an aerospace personal-equipment specialist, pulled Peterson clear of the wreckage. Peterson was put on a stretcher and loaded on the helicopter for transport to the base hospital.

One of the reasons Peterson survived the crash was a decision made during the construction of the M2-F2. Because of their shape, lifting bodies have tight center-of-gravity limits. It was necessary to put more weight in the vehicle's nose. No one now remembers who suggested it, but rather than just putting ballast in the nose, much of the added weight went into extra structure around the pilot. Rather than a 50-g cockpit, the M2-F2 had a 300-g cockpit, with a heavy steel frame around it. When the vehicle hit the lake bed and began rolling, the structure absorbed the energy and held it together, protecting Peterson.

EFFECTS OF PETERSON'S ACCIDENT ON THE LIFTING-BODY PROGRAM

Peterson's accident had the potential for a devastating blow to the lifting-body program. Less than six months before, the HL-10 had been grounded to sort out some major control problems during its first flight. We had flown two heavy-weight lifting bodies and had major problems with both of them. Were these lifting bodies really a viable lifting-entry spacecraft configuration, or were they destined to be as unflyable or as unpredictable as they looked?

There were many doubters before we flew them. Now there were more. The NASA headquarters managers were extremely concerned about the problems that we were encountering. It wouldn't take much to persuade them to give up on the program. They definitely did not want to kill anyone while flying these oddball shapes, and we had come close to doing just that in Peterson's crash. The results of the accident board's investigation could be critical in a decision on whether to proceed with the program or to give up.

The accident investigation board put most of the blame for the accident on the poor flying qualities of the M2-F2 and on management for the decision to continue flying even with the known PIO risk at low angles of attack. The board believed that the M2-F2 should have been modified to eliminate the PIO problem regardless of whether we had procedures to avoid it or techniques to stop the PIO if we did encounter it. They concluded that Peterson was disoriented as a result of his PIO encounter and this compromised his ability to make a successful landing.

That conclusion singled out the M2-F2 as the cause and absolved the pilot of any blame. The board's conclusion may have absolved Peterson, but it did him an injustice in concluding that he became disoriented because of the PIO encounter. During the PIO there were only two severe rolling oscillations before the M2-F2 stabilized and continued the landing approach. There was a 14-second period of stabilized flight from the completion of the PIO until the vehicle hit the lake bed. Peterson was a fighter pilot. Two quick rolls should not have disoriented him. Even if they did, there was the 14-second period to recover from any disorientation. In my opinion, the board was being protective of Peterson, but in doing so it did him a disservice.

When Bikle read the report, he asked me to review it and talk to the board chairman, Don Bellman. Bikle was afraid that the report might precipitate a cancellation of the program.

In reviewing the report and examining the events leading up to the accident, I came up with some other conclusions. In my opinion, the accident was a result of the near collision of the M2-F2 and the rescue helicopter. In the radio tape comments, Peterson's attention was fixed on the helicopter in the last seconds prior to the accident. His statements were, "Get that chopper out of the way" (9:41:50.1), "Move . . . chopper . . . goes . . . fast [possibly "west"]" (9:41:52.5), and "That chopper's going to get me, I'm afraid" (9:41:59.9). Impact occurred about 4 seconds later, at 9:42:04.2. The cockpit camera revealed that Peterson did not pull the gear handle until after the M2-F2 initially touched down. His attention was apparently riveted on the helicopter.

Two seemingly innocuous decisions were made that set the stage for the accident. The first was made the day before the flight and was to move the rescue helicopter from a hover position over the intersection of runways 18 and 23 to a position alongside 18, the intended landing runway.

For my first flights, I had intentionally positioned the helicopter over the runway intersection at 1,000 feet above the surface to serve as an altitude check to assist me in initiating the landing flare. When the helicopter appeared to pass through the horizon, I was 1,000 feet above the lake bed. I also positioned it there to make sure that it was out of the way and would not limit my maneuvering. This same position was used for all subsequent flights.

At the preflight briefing on May 9, 1967, the day before the mission, Gentry and Peterson decided that they did not need such an altitude reference. Therefore it was decided to move the helicopter to the same position used for an X-15 landing. This position was not clarified, however. Stan Butchart, the new director of flight operations, approved the change. Butchart had not been involved in the early decision to locate the helicopter at the runway intersection and thus was not aware of the reasoning behind that original decision.

The X-15 helicopter position was generally understood to be 1,000 to 2,000 feet east of Runway 18, but opinions differed on its north-south position. The flight controllers understood it to be 1.5 miles south of the north edge of the lake bed, and that was where the helicopter was located for Flight M-16-24. Peterson, however, understood it to be at the north edge of the lake bed. During X-15 flights, the helicopter was free to move around at will on the east side of Runway 18 and was only required to rendezvous with the X-15 after completion of its slideout.

The second decision was one made by Peterson during the flight to land across Runway 18 to align with a light crosswind. Peterson planned to touch down to the east of Runway 18 and roll out across the runway from east to west. No one noticed the potential conflict between the helicopter's position and Peterson's intended flight path.

Peterson's decision to land across the runway created another problem—a lack of depth perception. By landing east of the marked runway, Peterson would not have any good reference lines to aid in judging height. The runway lines on the lake bed were good references for judging height since we knew how wide they were. In many areas of the lake bed, there was no texture to the surface. Pilots could not judge height in these areas.

During the flight, Peterson encountered the PIO on final approach to his intended cross-runway landing at about 2,000 feet above the lake bed. The vehicle rolled violently through two oscillations and then stabilized. His intended landing location and heading may have been compromised by the PIO, but he completed the flare successfully and was lined up to land across the runway as he had intended.

It was at this point, having recovered from the PIO and still 14 seconds from landing, that Peterson saw the helicopter looming up directly in front of him. Worse, he never thought it would even be over the lake bed. He became fixated on the helicopter. His radio calls were all related to a potential collision with the helicopter. To make a good landing in the areas with no surface texture, the pilot had to concentrate all his attention on the lake bed surface. Peterson was not concentrating on the lake bed surface just prior to touchdown. Based on the radio conversations, he was concentrating on the helicopter.

The over-the-shoulder camera in the M2-F2 cockpit showed a steady slow rate of descent following the flare until the vehicle touched down. The cockpit camera then showed a slight lurch of the M2-F2, indicating a touchdown, just before Peterson pulled the landing gear handle. The ground camera on the lake bed showed that the landing gear deployment began following the initial touchdown. Touchdown occurred at 227.7 knots, about 10 knots higher than the normal landing gear deployment speed. The vehicle bounced into the air, recontacted the lake bed, rolled and tumbled, then slammed down. Years later, this accident scene was used to introduce the weekly episodes of the *Six-Million-Dollar Man* TV series. It was an impressive accident sequence. Anyone watching it would assume that the pilot had been killed.

Surprisingly, the pilot had only minor physical injuries. He broke a bone in his right hand and lost skin from his forehead and right eye area. Peterson was instrumented on this flight to measure such things as heart rate, breathing rate, blood pressure, and body acceleration. The instrumentation revealed that he was subjected to less than 5 g throughout the entire crash sequence. All of the vehicle's energy was dissipated in the rolling and tumbling motions. He might have survived the accident injury-free if the vehicle had not slammed on its last tumble.

It was assumed that Peterson's head was slammed into the lake bed on this final motion and that resulted in the loss of flesh on his cheek and forehead. Following the accident, Peterson was flown immediately to the Edwards hospital and then a few hours later was moved to the hospital at March Air Force Base because of the damage to the area around his eye. His eyelid had been torn away. He was finally transferred to the University of California in Los Angeles for specialized treatment for his eye injury. The eyeball itself was not seriously damaged. He could see, but the loss of the eyelid would ultimately result in damage to the eyeball unless an eyelid could be reconstructed. During the lengthy process of healing and preparation to repair the damage, Peterson contracted a staph infection, which eventually resulted in loss of vision in that eye. He currently wears an eyepatch.

Peterson came back to work eighteen months after the accident and began flying as a test pilot on a restricted basis. He obtained an FAA waiver that allowed him to fly with sight in only one eye. (Another one-eyed pilot was Wiley Post, who flew the *Winnie Mae* around the world in 1933.) Peterson also was allowed to continue flying in the U.S. Marine Corps Reserve as a helicopter pilot. He did surprisingly well flying with only one eye, but he was finally removed from piloting duty by NASA. He was appointed a program manager and finally became the Dryden director of system safety, reliability, and quality assurance before retiring from NASA in 1981. In 1987, he retired as a colonel from the Marine Corps Reserve. Following his retirement from NASA, he joined Northrop, where he retired as a system safety engineer with the B-2 program after eleven years.

I presented the results of my investigation of the accident to Paul Bikle and to the chairman of the accident board. The chairman of the accident board did not agree with all of my conclusions, but he did agree that the potential collision was a major factor. Bikle scheduled a meeting with the lifting-body program managers at NASA headquarters. He and I flew back to present a different version of the accident from that documented in the accident report. It took a lot of persuasion, but we finally convinced the program managers, Milton Ames and E. O. Pearson, that we should continue with the program.

The fate of the M2-F2 was not considered critical to the program. The program could continue without it. In fact, the M2-F2 and the HL-10 were each originally intended to be a backup for the other vehicle if either should be lost in an accident. The program was only attempting to demonstrate that a representative lifting body could be successfully maneuvered to a specific area and landed successfully following an entry from orbit. Either shape would accomplish this objective. Both Bikle and I, however, were interested in repairing the M2-F2. It was our first choice for a flight vehicle, and we were reluctant to continue without it. It was not our first choice because of its predicted flying qualities, though. In fact, Bikle told Al Eggers, one of the fathers of the M-2, that we wanted to fly the M-2 because it had the *worst* flying qualities. Bikle said, "If we can fly the M-2, we can fly any of the other lifting bodies." Eggers was at a loss for words when Bikle told him that, but he finally said that he didn't care why we wanted to fly it, as long as we flew it first. That we did.

Bikle's other concern about abandoning the M2-F2 was that we might have another accident and wind up with nothing to fly. It was still early in the program. We had not flown the rocket engine, nor had we gone supersonic. We still had a lot to do to document the flying characteristics of the lifting bodies. We needed a backup vehicle for the HL-10.

In retrospect, there was a bigger reason for wanting to rebuild the M2-F2. The years 1966 and 1967 saw a series of tragedies and setbacks at Dryden. Joe Walker had been killed in June 1966 in the XB-70 crash, December 1966 had seen the tip fin problems with the HL-10, and then the crash of the M2-F2 in May 1967. In November 1967, Mike Adams was killed in the crash of the X-15 number 3. The program continued for another year, but its fate was sealed with Adams's death. When it was formally ended in December 1968, the Advanced Vehicle Research Section's weekly report marked its passage by saying, "We are all rather sad to see those poor tired black birds retired to museums, but are even sadder when we look to the future for the next program on the horizon—and find nothing." For the first time since the X-1 arrived at Edwards in 1946, there was no groundbreaking, high-speed research aircraft program under way. The rebuilding of the M2-F2 would not only provide a backup should the HL-10 suffer a similar fate, but it would also start the rebuilding after all those setbacks. Paul Bikle told Bill Dana that we were going to rebuild the M2-F2 because we were not going to quit on a note of failure. To abandon the M2-F2 would be to admit defeat; to rebuild it was a reaffirmation of the future.

REBUILDING THE M2-F2

Bikle assigned me the task of justifying the repair or rebuilding of the M2-F2. At the time, I was director of research projects, and the lifting bodies were among the projects under my jurisdiction. The M2-F2's exterior was badly damaged in the accident. A cursory inspection immediately following the accident revealed extensive internal damage as well. We decided to send the vehicle back to Northrop for a detailed inspection of the structure to see if we could salvage anything.

The results of that inspection were even more pessimistic. The external surface of the vehicle was extensively damaged, except for the upper surface, which was protected by the vertical fins during the rolling and tumbling motions. As the vehicle was disassembled, it became evident that the internal damage was much more than superficial. Substantial damage to longerons, transverse bulkheads, and other secondary structures was noted. As the disassembly proceeded, damage to the primary structure was also evident. It soon became obvious that reconstruction would start almost from scratch. The entire structure appeared to be irreparably damaged.

That assessment was extremely demoralizing. We did not have the money to pay Northrop to build a new vehicle. We did, however, have a crew of mechanics and technicians who were unemployed as a result of the accident. John McTigue, the lifting-body project manager, asked Paul Bikle to give the team a chance to rebuild it. Bikle wanted very much to rebuild the vehicle, but the damage was severe, and it did not seem possible to make it flyable again. It took some strong arguments and arm twisting by McTigue and the team, but they finally got the go-ahead.

Although Bikle had given permission to rebuild the vehicle, it was done unofficially, as we had not yet persuaded the program managers at NASA headquarters to approve such an effort. Bikle and McTigue were not about to tell headquarters that the vehicle was not reparable. They directed Northrop to disassemble completely the M2-F2 as part of the detailed "inspection." They then directed Northrop to rebuild any parts that were damaged and begin reassembling the basic structure with the new parts.

The vehicle was stripped down to its two main longitudinal keel beams and placed in an inspection jig to check its alignment before the rebuilding began. McTigue and Bikle told headquarters that we were simply repairing the damaged parts as we reassembled them. The primary structure was rebuilt and assembled at Northrop using mostly new parts by a combination of Northrop and NASA personnel under the direction of Meryl DeGeer, the operations engineer.

While the vehicle was being disassembled, I began putting together a justification for repairing and rebuilding it. I had to hurry because our project personnel weren't waiting for approval to begin rebuilding it. I worked with Fred DeMerritte in the headquarters program office to justify the rebuilding. He was as interested in rebuilding it as we were. He could supply arguments and other material to me to use in my justification to his bosses. It was a cozy arrangement to have him as a mole in the program office. I sometimes used him to try out an argument on his bosses before using it in my justification.

The basic argument for rebuilding was the need for a backup in case of an accident or a major aerodynamic flaw in the HL-10 configuration that might ground it permanently. The HL-10 tip fin flaw had not been completely defined yet. The vehicle could have another flaw that might show up at transonic or supersonic speeds. The transonic characteristics of these vehicles were highly questionable. Few wind tunnels provided good transonic data, and fewer still could characterize these stubby nonaerodynamic shapes. We needed all the flight data that we could get to verify the controllability of these shapes in transonic speed regimes.

A second argument was that the M2-F2 probably defined the lower limits of subsonic performance for unpowered landings. With a maximum subsonic lift-to-drag ratio of 3 to 1, the vehicle was descending on a steep flight path: minus 27 to minus 30 degrees. Based on simulations and analysis, it appeared that a vehicle could not complete a flare and come level if its lift-to-drag ratio was much lower than 3. It didn't matter how much speed it had built up before the flare—it could not turn the corner and come level.

A more technical reason to justify the rebuilding was a proposal to develop and demonstrate a more advanced control system, which would be more representative of a lifting-entry spacecraft. The control system was to include advanced control laws, some elements of fly-by-wire, and blended aerodynamic and reaction controls.

As the justification was documented, it was circulated to the other research centers for comment and additional supporting material. When all the arguments for rebuilding were finally gathered and included in the justification, the package was submitted to NASA headquarters.

On March 22, 1968, a full ten months after the accident, we were given permission to begin the rebuilding. The task of persuading headquarters to rebuild the M2-F2 was an exhausting one, requiring many letters, phone calls, and meetings, but it was worth it in terms of team morale. I personally believe that the vehicle would have been rebuilt whether or not headquarters approved. NASA was lucky, in my opinion, to have approved it before it became a fait accompli. The rebuilding effort was well under way before we received formal approval.

The headquarters approval directed that the M2-F2's primary structure was to be partially rebuilt to allow it to be removed from the jig. The completion of the inspection was expected to take about sixty days. By late summer, the vehicle was to be shipped back to Dryden pending the headquarters decision to complete the reconstruction. This final go-ahead was given on January 28, 1969. Northrop continued to fabricate new parts, which were then installed by the NASA team at Dryden. Bikle did this to give our mechanics and technicians the experience and also to minimize the visible dollar costs. In-house manpower costs were not accounted for in that era. Even so, the costs of repairing the vehicle were high, since we were buying all new parts.

The overall rebuilding process was slow. It took two and a half years to rebuild the vehicle and then prepare it for its next flight. We also made some significant modifications during reconstruction to improve pilot safety, the rollover structure, and its flying qualities. The only obvious modification was the third vertical fin, located between the two outboard fins. This was intended to cure the PIO problem.

The PIO was caused by deflecting one of the upper flap segments to roll the vehicle. This created a high-pressure region above the deflected flap, which created an asymmetric force on the inboard side of the vertical fin, causing the vehicle to yaw rather than roll. The ultimate solution was to install a center fin between the two upper flap segments. This eliminated the yawing moment produced by the upper flap deflection, since the pressure over the flap would act equally on the center and outboard vertical fins, canceling out the yaw moment. Because of the modifications, the vehicle was renamed the M2-F3.

FIXING THE HL-10

While we were awaiting official approval to rebuild the M2-F2, we were also trying to fix the HL-10's flow separation problems. During the winter and spring of 1967, Langley had come up with two different fixes. Modification 1 involved a thickening of the inside surface of the tip fins, and modification 2 proposed changing the shape of the leading edge of the tip fins. Langley ran a full set of wind tunnel tests on each and provided the data to Dryden.

Bob Kempel spent the summer of 1967 reviewing the data. He began plotting the data, literally thousands of points, by hand. The computer data plotting programs we now take for granted did not exist. The data for both modifications were plotted as a function of angle of attack for constant Mach numbers. The scale was uniform to allow easier comparisons. Initially, there seemed to be little difference between the new tip fin designs and the original data. On closer examination, there were subtle but significant differences. Some nonlinearities present in the original data were not there in the new plots. Kempel assumed that if the original nonlinearities indicated flow separation, then their absence would indicate no flow separation or a lesser amount. Based on this, Kempel recommended modification 2.

In early autumn 1967, NASA contracted with Northrop to make the modifications. They were relatively simple—a fiberglass glove backed with a metal structure would be attached to each tip fin leading edge. Work continued throughout the winter. The NASA-Northrop teamwork that had produced the original construction continued. At the NASA hangar, Fred Erb, a senior engineer with 25 years experience with Northrop, shed his suit, put on overalls, and pitched in with the rest of the team.

With Peterson injured in the M2-F2 crash, a new pilot was needed for the second HL-10 flight. Jerry Gentry was selected as HL-10 project pilot. He had flown the M2-F2, experienced its PIO tendencies, and liked neither. During his simulator practice, it was apparent to the HL-10 team that he was skeptical. Gentry never said it out loud, but clearly he thought the simulator was too good. After many hours in the simulator, he was finally ready to fly.

The second glide flight of the HL-10 was made on March 15, 1968. Launch from the B-52 was made at 45,000 feet at Mach 0.65. As the vehicle descended toward the lake bed, Gentry made mild pitch and roll maneuvers at angles of attack up to 15 degrees to evaluate any control degradation, as on the first flight. A movie camera had been mounted in the center fin to view the right inboard tip fin flap and right elevon. These areas had been covered with dozens of 6-inch-long strands of wool yarn. Called tufting, this procedure had been used since the days of wood and fabric biplanes to allow aerodynamicists to see airflow patterns. If the airflow remained attached, the yarn would lie flat on the surface in the direction of the flow. If the airflow separated, the yarn would follow the disturbed flow by moving around randomly. Gentry also made a simulated 2-g flare at altitude to assess the landing characteristics. Flight H-2-5 lasted just short of 4.5 minutes and ended with a successful landing.

The flight was both a success and a relief for everyone concerned. Gentry found that the HL-10 handled as well as the F-104 flying a simulated approach. The longitudinal stick was a little sensitive but was still satisfactory. He recommended no changes for the next flight. The flight data and the camera film showed no trace of flow separation or control-system problems. The HL-10 flew as well as the simulator had indicated. It was far better than the M2-F2, and pilots were lining up to fly it. Most important, the lifting-body program had started its recovery from the hard times of the previous years.

JOHN MANKE'S INTRODUCTION TO FLIGHT RESEARCH

The next four glide flights were also made by Gentry. The seventh and eighth HL-10 flights, on May 28 and June 11, 1968, were made by NASA research pilot John Manke. Manke went on to make forty-two lifting-body flights, but when he joined the pilots' office years before, he had to go through an apprenticeship just like the rest of us. Each new research pilot had to prove himself, both in the air and on the ground. He was also subject to practical jokes. When Manke would come into the office, he would discover that Bruce Peterson and Bill Dana had moved his desk out in the hallway. He got no help from the rest of us and had to put it back himself. The next time he went out, the same thing would happen.

The new pilots also got the grunt jobs. One of these was flying the C-47, which meant taking people to LAX, picking up spare parts, or flying out to the ground stations. It was not popular duty, and Manke delayed his C-47 checkout. This did not go unnoticed. One morning he came into the pilots' office, which was quite a large room, and discovered that someone had turned it into a C-47 simulator. He found two rows of seats, with two large seats at the front for the pilot and copilot. Two large fans on poles stood in for the engines, and there was a pilot's manual and fake control yokes. Manke got the message: he had to get checked out.

Manke had an added burden in that he was a Marine pilot; the other NASA research pilots had Air Force or Navy backgrounds. The kidding this caused lasted well into the lifting-body program. After landing, the time and location of the postflight debriefing would be announced. Usually it was scheduled for an hour or so later, to give the pilot a chance to get out of the pressure suit, take a shower, and get dressed. After one of Manke's flights, Jerry Gentry (Ol' Silver Tongue) announced, "The postflight debriefing will be held at fourteen hundred hours." He paused, then added, "For you civilian pilots, that is two o'clock." He paused again, and concluded by saying, "And for you Marine pilots, Mickey's little hand is on the two."

DEVELOPMENT OF THE SV-5J

Even as the lifting-body program was threatened with cancellation by NASA, it was also expanding with the development of a third lifting-body shape by the Air Force. The SV-5 shape had gone through a long development. The program had started in August 1960 with a request by the Air Force's Air Materiel Command for development of a small maneuverable data return capsule for the SAMOS reconnaissance satellite. In November 1960, the Martin Company was selected to develop a lifting-body design. Over the next three years, the Air Force, the Aerospace Corporation, and Martin looked at several different lifting-body shapes, including two versions of the M-2.

The shape that finally emerged was developed by Hans Multhopp of Martin. During World War II, he had been on the design staff of Focke-Wulf, heading the aerodynamics department and later the advanced design department. Among his wartime designs was the Ta 193, the first swept-wing jet fighter.

John Rickey, a Martin aerodynamicist, was assigned to refine the basic SV-5 shape into a vehicle able to make both a hypersonic reentry and a low-speed landing. Low-speed wind tunnel tests showed a need for a minor change in the nose shape, as well as a center fin and special care in the shaping of the tip fins to prevent flow separation. (The latter two problems would appear in the M2-F2 and HL-10 test flights.) In December 1963, Martin finalized the SV-5 shape. At the same time, we were flying the M2-F1, and the Dyna-Soar was canceled.

The small unmanned SV-5D lifting body was to be test flown at suborbital speeds on an Atlas booster. It would maneuver during reentry, and then parachute to a landing. The SV-5 shape was, without question, the ugliest of the original lifting bodies. It resembled a potato with three fins. During the summer and fall of 1964, development work on the SV-5D began at Martin.

At Dryden, I was flying the M2-F1 and Northrop was building the M2-F2 and HL-10 vehicles. Our efforts were attracting attention and a shift in the Air Force lifting-body program. Interest grew in building a full-scale manned SV-5 for low-speed flight tests, similar to what NASA was planning for the M2-F2 and HL-10, and it was approved in January 1965. The SV-5D (later renamed the X-23A) was to test lifting-body aerodynamics at orbital speeds; the SV-5P (later renamed the X-24A) would provide data on transonic and supersonic handling qualities. As with the M2-F2 and HL-10, responsibility for the X-24A was split between the Air Force and NASA. Soon after the SV-5P/X-24A was approved, a third design was added to the SV-5 lifting-body program: the SV-5J, an attempt to build a jet-powered lifting body as a training vehicle for student test pilots.

Chuck Yeager was enamored with the lifting body as a training vehicle for the Aerospace Research Pilots School. Yeager had made several flights in the lightweight M2-F1 after I checked him out and was convinced that a lifting-body training vehicle would provide a demanding task for his new astronaut candidates. Martin was approached by Yeager to develop a jet-powered training vehicle. Buzz Hello, the X-24A program manager for Martin, was intrigued by the idea. He directed his people to look at modifying the X-24A to a jet-powered vehicle.

Northrop was also aware of Yeager's interest and looked at modifying the M2-F2 or HL-10 as a jet-powered trainer. Northrop, however, gave up on the jet-powered vehicle when it became obvious that the thrust required to fly a lifting body was quite substantial (even though it was a small vehicle and relatively light in weight). The low lift-to-drag ratio was the killer. The M2-F2 would require close to 3,000 pounds of installed thrust to achieve a takeoff. There were existing engines that would fit into the lifting bodies and provide the required thrust, but the installed thrust would be marginal with realistic inlet ducts. Excessively large inlets to obtain the needed mass flow would compromise the lifting-body shape and require a new set of wind tunnel tests to define their characteristics. Large inlets would also add drag, something a lifting body already had in excess. Northrop backed out of the competition to develop a jet-powered lifting body in part because of the limited market for such a vehicle. Yeager was interested in buying only two or three vehicles.

Martin, on the other hand, continued to pursue the idea. The company developed a jet-powered configuration that didn't look too bad, but it was still draggy and underpowered, using the best available engine that would fit within the body of their X-24A design. The fuselage was the same size and shape as the X-24A. (The X-23A, X-24A, and SV-5J all used the same shape, even though the X-23A was much smaller.) The only obvious difference between the

X-24A and the SV-5J was the underslung inlet under the SV-5J's fuselage for the J-60 engine. The design did offer enough hope to proceed with some engine tests, a vehicle simulation, and construction of two fuselage shells.

One day, the Martin Company on-site representative at Dryden came to my office to talk about the SV-5J program. He concluded the discussion by asking if I was interested in testing the vehicle for Martin. I said I might be interested, but I didn't know how I could do it while still an employee of NASA. I talked to Paul Bikle about the offer, and he immediately offered to give me a leave without pay to enable me to test the vehicle. I was quite surprised at how quickly he responded, because I hadn't made up my own mind on whether I wanted to do it. I had already quit flying at Dryden, and I would need to get proficient again.

I thought long and hard about the offer from Martin, but I wasn't too impressed by the low-cost approach that Martin was proposing to build and test the airplane. They planned to build it in Baltimore and then ship it out to Edwards for the flight program. Martin had no facilities at Edwards, but Yeager offered them some space in his Aerospace Research Pilots School hangar to check out the vehicle and prepare it for flight. Martin would send a minimum crew to Edwards and depend on Chuck Yeager to provide any additional support that they might need. Yeager would also allow me to fly his aircraft for proficiency purposes. I compared their approach with the approach that we had taken in getting the M2-F2 and HL-10 ready for flight, and I was hesitant to take them up on their offer. We had put a lot of manpower and effort into the preparation of our vehicles for flight. I wasn't convinced that they planned to do the same.

While I was pondering Martin's proposal, they offered to fly me back to Baltimore to talk to their senior management and also fly their simulator. I agreed. I was introduced to the pilot who had been flying their simulator and serving as project pilot. I didn't ask if he had been offered the job of testing the SV-5J. It didn't seem like the polite thing to do. He indoctrinated me into the simulator and pointed out some techniques that he had been using to make successful flights. One crucial task was to get the landing gear up immediately after takeoff before the vehicle settled back onto the runway. The vehicle was terribly underpowered. He could get airborne with the gear down, but once he began retracting the landing gear, he began losing airspeed because of the extra drag during the landing gear cycling process. If I didn't get the gear up at the proper time, I would be back on the runway with the gear partially retracted—not the way to make a leisurely first flight. I couldn't believe they were really serious about building and flying such a marginal vehicle.

After I finished my simulator evaluation, I was escorted upstairs to meet the vice president responsible for this particular project. He wanted my impression of the vehicle based on the results of my simulator evaluation. I was initially a little embarrassed about telling him what I really thought of the vehicle, but I finally began giving him my impressions and their implications. I told him that my major concern was the lack of adequate thrust to make a normal, safe takeoff. Depending on a quick gear retraction was not the way to guarantee a safe takeoff. I told him that even if I could do it safely, some of Yeager's students might not. An underpowered vehicle like that was a guaranteed killer. Martin could only ruin its own reputation by killing a couple of astronaut candidates.

I reminded him that we were going to fly the X-24A at Edwards, and based on what I knew about it, I believed it would fly safely and also fly well. It appeared to have good flying qualities based on my simulator evaluation. The X-24A might enhance Martin's reputation. The jet version would definitely degrade it. I concluded my evaluation of the SV-5J by strongly recommending that Martin not pursue the idea of flying it.

I returned to Edwards that evening and assumed that Martin would give up on flying the SV-5J. Several days later, I received a call from the Martin representative who had first contacted me about the SV-5J. He asked me to quote a price for testing the vehicle. I told him I wasn't interested. He persisted, but I finally convinced him that I really wasn't interested. I suggested that he talk to one of the other lifting-body pilots. The next day, I got another call asking me to quote a price. I again indicated I was not interested. A day later, another call. By this time, I had decided to quote them a ridiculous price to get them off my back. I said that I would make the first flight for $25,000. That doesn't sound like much money now, but at the time it was equivalent to a year's salary for me.

I also told Paul Bikle how I intended to make that flight. I planned to put a two-by-four board across the runway about 1.5 miles down the runway and then aim to hit it, causing the SV-5J to bounce into the air. If the vehicle began flying, I'd raise the gear and continue the flight. If the vehicle began to settle back down after hitting the two-by-four, I would just land it and ask for my $25,000. I didn't intend to specify any flight duration in the contract for the first flight. A flight was a flight, whether it was an hour or just 5 seconds in duration. I never did get a response to that offer. Martin gave up on flying the SV-5J, and the vehicle was quickly forgotten.

CONVINCING A PILOT TO FLY THE M2-F3

When the M2-F3 was finally repaired, Paul Bikle and I had to do some fence mending. The remaining lifting-body pilots still had a lot of hard feelings over the M2-F2 PIO and Peterson's crash. Bikle suggested I talk with them and convince one or the other that the new M2-F3 was now a good flying machine. John Manke and Bill Dana were the two remaining lifting-body pilots.

I invited them to my office, closed the door, and began a heart-to-heart talk about the M2-F3. They were both pretty blunt about the M-2. It had clearly demonstrated its tendency to PIO on at least four occasions before the crash. Don Sorlie, Jerry Gentry, Bruce Peterson, and myself had all experienced PIOs flying the M2-F2.

Manke and Dana faulted Paul Bikle and me for continuing to fly the M2-F2 with a known handling-quality deficiency. They believed we should have stopped flying the vehicle and fixed it before continuing to expand the flight envelope. They had a valid criticism. We were taking an unnecessary risk if time was not a factor.

My primary argument was that the PIO region was well defined, we had several positive ways of recovering from it, and each of the pilots had successfully recovered from his encounter. I also argued that we had flown other research aircraft with similar bad or even dangerous flight characteristics and had managed to do it safely. Sometimes a fix for problems of this type was not easy to define. I had to admit that I was in a hurry to get on with the program and that was part of the reason that we had not fixed the M2-F2.

What we had done or not done in the past, however, had little to do with where we were going from this point on. I argued that the modifications made to the M2-F3 (the addition of a center fin) had eliminated the PIO problem. This argument was based on new wind tunnel data and simulations. The vehicle now appeared to be a good flying machine.

The meeting broke up after about an hour with no one completely convinced of where we should go from here. Bill Dana did agree to get more involved in the simulation and control design effort. That was a major victory for me, because he had been (and still is) critical of the M2-F2. It was admittedly the worst of the lifting bodies in terms of performance and, with the PIO tendency, the worst in terms of flying qualities. A number of people considered it an accident waiting to happen. Dana used to say that the only thing worse than being pilot of the M2-F2 was to be copilot. The M2-F2 did not have a second seat, but I would have to agree with Dana. I'm quite certain that we would not have gotten a volunteer to fill that seat, if we had one.

DANA

As it turned out, Dana became the M2-F3 project pilot. Soon after he had been selected, Jack Kolf was looking at the "NASA" painted on the vehicle's fin and realized how much it looked like "DANA." He went to Billy Shuler at the Dryden paint shop to make a stencil. They then painted "DANA" on the M2-F3's new center fin. The black lettering on a yellow stripe was in exactly the same style as the "NASA" painted on the tails of the Dryden aircraft.

For two days the vehicle sat in the hangar without anyone noticing that the "D" and "N" had replaced the "N" and "S." Finally, Paul Bikle was walking through the hangar. He went past the M2-F3, then stopped and did a classic double take. He stared at Kolf and Shuler's handiwork, then shook his head and walked on without saying a word.

Even with the hard times, the people working on the lifting bodies never lost their sense of humor. The work was hard and the hours were long, but it was a fun place to work. Many of the people who worked on the lifting-body programs said those were the best years of their lives. We were a small group of young, eager individuals facing a difficult, demanding task who became not just a team, but a big family. We were in the middle of a desert, far from headquarters and from headquarters attitudes. Each lifting body had a crew of about thirty people, so everyone knew everyone else. Lifting-body personnel would also be switched around from project to project as needed. If someone needed help, a tool, or a part, they would pitch in to help.

By 1969 and 1970, the hard times were over; the HL-10 had been fixed, the X-24A had joined the program, and both were flying. The M2-F3 had been repaired, and it, too, would soon fly. We had reconstructed and tamed the savage beast, even if it still had claws. The lifting bodies were now ready to push the envelope.

9
PUSHING THE ENVELOPE

This chapter was written by Curtis Peebles using Milt Thompson's outline along with published accounts and documents.

The goal of the heavyweight lifting-body program was to test the vehicles at transonic and supersonic speeds. The M2-F2 had been on the verge of its first powered flight when it was nearly destroyed in the crash. The tip fin problems with the HL-10 delayed its glide flight tests for fifteen months. It was not until the spring of 1968 that the first glide flights were made by Jerry Gentry and John Manke. Between March and October they made a total of ten glide flights to test the tip fin modification. Of the original lifting bodies, the HL-10 proved to have the best flight characteristics, and it was time to make a powered flight.

THE HL-10'S PUSH TO SUPERSONIC SPEEDS

The first attempt was made on October 23, 1968. The launch at 40,000 feet was successful, but Gentry couldn't get one of the rocket chambers to ignite. He made several attempts, but nothing worked. Gentry finally shut down the engine. As the vehicle passed through 25,000 feet, the pilot jettisoned the remaining fuel. At that point, the stability control system malfunctioned, causing several oscillations. Gentry made adjustments to the flight controls and continued toward a landing. Because of the engine problems, Gentry was headed toward Rosamond Dry Lake, to the west of the Edwards lake bed. He touched down successfully on the empty lake bed. He then had to wait 20 minutes for the ground crew to show up and help him out of the vehicle.

The first successful powered lifting-body flight was made on November 13, 1968, on the HL-10's thirteenth flight. John Manke fired two of the engine chambers and reached Mach 0.84. The vehicle then began a speed and altitude buildup. The three lifting bodies—the HL-10, X-24A, and M2-F3—had radically different shapes, but each would undergo a similar set of tests. This started with glide flights, followed by a buildup to supersonic speeds, aerodynamic tests, and finally maximum speed and altitude flights. By this procedure, the three shapes could be directly compared with each other. In effect, it was a flyoff, much as the Air Force might test different prototype fighters before awarding a production contract.

Gentry made his last HL-10 flight on December 9, 1968, a two-chamber flight to Mach 0.87. He then transferred to the X-24A program as project pilot. For his work on the HL-10, Gentry was awarded the Harmon International Trophy for the most outstanding contribution to the science of flying.

On April 17, 1969, Manke reached a speed of Mach 0.99 using three chambers of the HL-10's engine. Manke and the HL-10 team began a month of preparations for the first supersonic lifting-body flight. They started with a complete review of the wind tunnel data. The analysis predicted acceptable levels of longitudinal and lateral-directional dynamic stability at all Mach numbers and angles of attack. This was true even if the static directional stability was zero or slightly negative, providing that the angle of attack was a positive value. To demonstrate this, a special, fictional, simulator data set was developed with the static directional stability set to zero. The simulator crew demonstrated to Manke that even under those conditions the vehicle would remain stable.

One day Manke was in the HL-10 simulator during the lunch hour, brown bag beside him. The simulator engineer was the only other person present. He loaded the data set into the computer, then left for his own lunch. Manke could run the simulator on his own, by pushing the "operate" and "reset" buttons. What he did not know was that the data set loaded into the computer was the fictional one with the zero static directional stability values, not the real one used for training. Not realizing the error, Manke began the simulator run for the supersonic profile. Reaching the planned altitude for the acceleration to

Mach 1, he pushed the nose over, and the vehicle became violently unstable in a lateral direction. The flight plan called for an angle of attack of 6 degrees, but he had inadvertently approached 0 degrees. He crashed.

Manke expressed his intense dissatisfaction with the simulation to NASA management before reporting the problem to the project engineers, who were still out to lunch. Before the project engineers knew what had happened, Garry Layton, the lifting-body program manager, Wen Painter, Berwin Kock, and Bob Kempel were summoned to the Bikle barrel, one of the executive offices. Paul Bikle, director of research Joe Weil, and his deputy Jack Fischel asked them to explain why they were trying to kill a perfectly good research pilot, even if he was from South Dakota. After the feeding frenzy was over, the project engineers were given their say. It finally dawned on them what had happened, and they explained the problem. This was followed with a demonstration in the simulator. With the correct data set loaded into the computer, no dynamic instability occurred.

The first supersonic lifting-body flight was made on May 9, 1969. The flight plan was for Manke to be launched from the B-52 about 30 miles from Edwards. The pilot would then ignite three rocket chambers, rotate to a 20-degree angle of attack, and maintain this until a pitch angle of 40 degrees was established. This pitch angle would be maintained until an altitude of 50,000 feet was reached. The pilot would then push over to a 6-degree angle of attack and accelerate to a planned Mach number of 1.08. Manke would then turn off one rocket chamber and maintain a constant Mach number while gathering data. A typical 360-degree approach would end with a landing on Runway 18.

The flight itself was uneventful. A top speed of Mach 1.127 was achieved, along with a maximum altitude of 53,300 feet. Manke said that "everything went real well." Years later, several of the project engineers thought the events leading up to the flight were more exciting than the actual flight. That aside, the first supersonic lifting-body flight was a milestone, both aerodynamically and psychologically.

The vehicle showed acceptable transonic and supersonic handling characteristics, and at subsonic speeds it handled better than some existing fighters. The only surprise involved a transonic pitch trim change that occurred between Mach 0.96 and 0.97. It had been predicted, but the onset was more abrupt than expected. It was apparently caused by an abrupt flow change or shock-wave movement. On another level, reaching supersonic speeds represented a psychological achievement as well. How an aircraft would react to Mach 1 was always a question, one of even greater concern for the lifting bodies because of their early problems.

During the spring and summer, Bill Dana and Air Force major Peter Hoag joined the HL-10 program. After completing ground school and simulator training, they each made a glide flight in the HL-10. Dana completed his on April 25, 1969; Hoag followed on June 6. Each pilot then made a subsonic powered flight for pilot familiarization.

Throughout the summer and fall of 1969, the HL-10 was flown to ever higher speeds and altitudes, expanding the vehicle's envelope. Following the first supersonic flight on May 9, which reached Mach 1.127 and 53,300 feet, Manke raised those numbers to Mach 1.236 and 62,200 feet on May 28. His June 19 flight reached Mach 1.398 and 64,100 feet, and on August 6, he achieved Mach 1.54 and 76,100 feet on the first four-rocket-chamber flight. Bill Dana surpassed that altitude on September 3, reaching 77,960 feet. Manke's last HL-10 flight, on September 18, reached 79,190 feet. He then left the HL-10 program to join Jerry Gentry on the X-24A program, which was then conducting glide flights.

DANA'S PINK SPACE BOOTS

Because powered flights would exceed an altitude of 50,000 feet, a pressure suit was required should cockpit pressurization be lost. For the lifting-body flights, each pilot wore a custom-tailored A/P-22S-2 pressure suit. Like the suits for the X-15, Dyna-Soar, SR-71, and Gemini, these were made by the David Clark Company. The contract with NASA said that the A/P-22S-2 suits for the lifting-body pilots would be white with black boots. The Air Force contract for the same suits for SR-71 crewmen specified a white suit with white boots.

When Bill Dana went to the fitting, he found his suit had white boots. He thought they looked sissy and said that he might as well fly in pink boots. The people at David Clark reassured him that it was a mistake and they would correct the problem. When Dana's suit was delivered, the boots were black. David Clark also included a spare pair of space boots. They were pink with adhesive yellow flowers stuck on them. Dana wore the pink boots on his next HL-10 flight. After landing, he went to the pressure-suit van to change. When he came into the postflight debriefing, he climbed up on the table wearing the pink boots to show everyone he had actually worn them.

THE FASTEST AND HIGHEST LIFTING BODY

Following the envelope expansion, the HL-10 pilots made a series of flights in late 1969 to collect data on stability and control at different angles of attack, performance, and stability augmentation settings. The increases in speed and altitude of these flights were incremental. On October 27, Dana reached a speed of Mach 1.58, and on November 17 he raised it to Mach 1.59. Those speeds were only Mach 0.04 and Mach 0.05 faster than Manke's August 6 flight. The altitude of the flights showed a similarly small incremental increase. On November 21, Peter Hoag flew the vehicle to 79,280 feet, a mere 90 feet higher than Manke had reached on his last HL-10 flight. Dana got to 79,960 feet on December 12, 1969, on the vehicle's thirty-first free flight.

With the new year of 1970, the HL-10 team embarked on a final series of powered flights to gain stability and control data close to the vehicle's speed and altitude limits. The series began on January 19, 1970, when Hoag reached 86,660 feet. A week later Dana flew to 87,684 feet.

The HL-10's final two powered flights were to be the fastest and highest made by any of the lifting bodies. On February 18, 1970, Hoag made the fastest lifting-body flight. The flight plan called for a launch at 47,000 feet, higher than the normal 45,000-foot drop altitude. Immediately after launch, the pilot would ignite all four rocket chambers, then establish a 23-degree angle of attack until a pitch angle of 55 degrees was established. This would be maintained until an altitude of 58,000 feet was reached. Then the pilot would make a zero-g pushover until an angle of attack of 0 degrees was reached. This would be maintained until the fuel was exhausted.

The flight went as planned, reaching a peak speed of Mach 1.861 at 67,310 feet. The goal of the flight was to investigate suspected roll reversal at high speed and to operate the vehicle for an extended time with the stability augmentation system off. Following the successful lake bed landing, the preliminary analysis indicated that no roll reversal actually existed, and that the vehicle's handling was better than anticipated. The pilot said that it had handled "beautifully" in spite of moderate air turbulence.

Dana made the HL-10's final powered flight to maximum altitude on February 27, 1970. Launch would be at 45,000 feet, followed by ignition of all four chambers and a 23-degree angle of attack to a pitch angle of 55 degrees. This would be maintained until Dana reached an altitude of 75,000 feet. He would then perform a pushover to an angle of attack of 7 degrees, which would be held until Mach 1.15 was reached. Then he would open the speed brakes, which would increase the angle of attack to 15 degrees. The flight was to obtain data with the speed brakes open at supersonic speeds and to obtain stability and control data at supersonic speeds.

The peak altitude reached was 90,303 feet, the highest altitude any of the original lifting bodies would ever achieve. The flight went as planned, except that a high crosswind moved the landing over to Runway 23, east of the usual runway, 18. Total time from launch to landing was just under 7 minutes.

Following Dana's altitude flight, the HL-10 was modified for the final two flights, 36 and 37. The vehicle undertook tests of a powered approach and landing in support of the emerging space shuttle program.

With the tip fin modifications made after the first glide flight, the HL-10 was universally praised by its pilots as the best handling of the original lifting bodies. Among Dryden pilots and engineers, it was the preferred shape for the space shuttle. There were no serious incidents involving the HL-10 either in flight or on the ground.

The only real problem with the HL-10 shape was its Plexiglas nose window, which was required for pilot visibility during landing (a feature the M2-F2/F3 also shared). On the HL-10, this provided excellent visibility, but the clear nose was lenticular in shape, which acted like a demagnifying lens. When the vehicle was close to the ground, it gave the pilot the impression he was higher than he was. Manke reported after one flight, "I touched down before I wanted to." Several of the pilots, on their initial flights, waited until they were very close to the ground before extending the landing gear. This problem was eased with pilot experience.

THE X-24A GLIDE FLIGHTS

Martin began construction of the X-24A following an initial design review in April 1966. Work was completed in the late summer of 1967, and the Air Force accepted the vehicle at Martin's Middle River plant near Baltimore on August 3, 1967. Martin then removed the fins and loaded the vehicle on pallets for transportation to Edwards. The X-24A arrived aboard a C-130 cargo plane on August 27, 1967.

The delivery took place in the midst of the hard times. Paradoxically, the X-24A gained from the painful experience. The HL-10 tip fin problems and the M2-F2 control problems were both addressed early in the design phase. The

X-24A was also farther along in the evolutionary process than the first two lifting bodies, resulting in a more practical design. The vehicle's forward cockpit location with a bubble canopy improved pilot visibility during landings.

Johnny Armstrong was named Air Force program manager for the X-24A program. Piloting and engineering duties would continue to be shared between NASA and the Air Force, but the Air Force engineering team would concentrate on the X-24A and have primary responsibility for documentation of the test results on that vehicle. That put the Air Force engineering team in an unusual position. Dryden was responsible for lifting-body flight safety, so the Air Force Flight Test Center management paid little attention to the team's day-to-day activities. At Dryden, the NASA engineering and maintenance personnel viewed the Air Force team as the final authority on all activities regarding the X-24A, even though the team was outside the NASA chain of command. (The X-24A had originated with the Air Force and was managed and paid for totally by the Air Force.) The result of this nonstandard management arrangement was an unusually strong sense of personal responsibility, high morale, and high job satisfaction. The Air Force engineering team checked and rechecked their results and recommendations before submitting them to Dryden. It was also possible to deal with problems much faster within the team than through the normal NASA or Air Force chain of command.

Following the X-24A's arrival at Edwards, the team began installing and calibrating the research instrumentation. This required that the cockpit be completely dismantled. Unlike Northrop, Martin was a closed union shop, so the NASA engineers did not have the same freedom to work on the vehicle during construction that they had with the Northrop M2-F2 and HL-10. (Northrop was a nonunion shop.) This was true even when they knew modifications would have to be made later. By February 1968, the installation of the instrumentation and other equipment was complete. The vehicle was readied for shipping to Ames for tests in the 40-by-80-foot wind tunnel.

To avoid the problems of the M2-F2 and HL-10, the data from the small model wind tunnel tests were reexamined. Additional model tests were run to fill in gaps in the data. The X-24A was also tested in the full-sized wind tunnel, which confirmed the stability and performance estimates from the model tests.

Another test done in the large tunnel measured the effects of a simulated ablative heat shield on vehicle performance. All the heavyweight lifting bodies were built of aluminum, as they would not be exposed to significant aerodynamic heating. To withstand the heat of actual reentry, that smooth aluminum skin would have to be covered by an ablative heat shield made of plastic. As it was exposed to reentry heating, the plastic material would melt, carrying away the heat and leaving a charred, rough surface.

To simulate such a shield, a water-soluble glue was sprayed over the vehicle, and a wire mesh was laid over the surface to simulate the honeycomb pattern of the ablative material. Coarse sand was then sprayed on the glue. Once the glue had dried, the wire mesh was removed, leaving a rough pattern on the surface. The plan was to correlate the data from the X-24A wind tunnel tests with the results from the X-23A suborbital flights. The X-24A tests indicated a reduction in the lift-to-drag ratio of about 20 percent and a loss of stability due to the roughness of the simulated heat shield. Any plans to test the vehicle with the simulated heat shield in flight were dropped.

Once the wind tunnel tests were completed, the X-24A was returned to Dryden for checkout. The experience with the M2-F2 and HL-10 enabled the engineers to identify possible problems with the control system. These ground tests identified a large dead band in the pitch axis (2 degrees of flap travel) when control was being transferred from the lower flap to the upper flap. The ground test also indicated that if the controls were maintained in this dead band for more than 3 seconds, a divergent vibration was possible.

Tests with the X-24A simulator also showed problems with the complex control laws developed by Martin. This was an automatic program that would change the upper flap position, rudder position, and elevator gearing as the Mach number changed. The dead band was a concern, as was the lack of a backup to the automatic program and the inability to make a practice flare at altitude. The Martin control laws were judged unsuitable for the early X-24A flights, and a simpler program was developed in which the pilot could manually flip a switch. This switch symmetrically and simultaneously changed the upper and lower flap positions and the stick gearing. The X-24A, like the M2-F2, also had an aileron-to-rudder interconnect that could be changed manually by the pilot.

By March 1969, the vehicle was ready for taxi tests using the landing rockets. On March 4 and on March 11 through 13, a total of twelve taxi runs were made. The X-24A was the first lifting body to be equipped with a steerable nose gear. The nose gear steering was from a T-39 business jet and was designed for low-speed operation, not the 200-plus knots of an X-24A rollout. This meant that it was much too sensitive. A steering failure could lead to a rollover accident, much like the crash of the M2-F2, so it was disconnected. Following an unmanned B-52 taxi test on April 2 and a captive flight with Jerry Gentry aboard two days later, the X-24A was ready to begin glide flights.

The first X-24A glide flight was made on April 17, 1969. Gentry made a successful practice flare at altitude, but he noticed some lateral sensitivity at high speeds during the final approach. It reminded him of the unstable feeling of the M2-F2 under the same conditions. The pilot decided to slow down from 300 knots to 270 knots before beginning the flare and fire the landing rockets at the start of the flare. As the gear deployed, there was a pitch down of the vehicle, as had been expected (and which occurred with the other lifting bodies). A normal landing was made. The postflight analysis indicated that a malfunction had occurred in the rudder-to-aileron interconnect electronics, and that was considered to be the cause of the lateral sensitivity.

Later that same day, John Manke made a powered flight to Mach 0.99 in the HL-10 (the only time two lifting bodies flew on the same day). The different vehicles were being tested on a staggered schedule: while the HL-10 was beginning its powered flights, the X-24A was making its initial glide flights; as the HL-10 program was nearing its end, the X-24A would be undertaking powered flights, and the M2-F3 would be starting its first glide flights.

The second X-24A glide flight followed on May 8 (the day before the first supersonic HL-10 flight). This time the interconnect worked correctly, but Gentry again detected lateral sensitivity and rolling motions on final approach. The pilot again slowed down and fired the landing rockets during the flare and landing.

The results of the first two X-24A glide flights were disquieting. Like the M2-F2 and the HL-10, the X-24A seemed to show stability problems. A series of reviews were begun by both NASA and Air Force safety groups, which were cross-checked by the individual engineers with the lifting-body program. These continued into the late summer of 1969. The cause of the lateral stability problem was traced to a combination of errors.

A significant error existed in the wind tunnel predictions of the lower flaps' effectiveness in yaw. Apparently, the error had been caused by flow interference between the support mounts and the lower flaps. Based on the erroneous data, the control settings had been incorrectly programmed. Several changes in both the mechanical and electronic systems corrected the problem. The Air Force engineering team felt that the rolling motions on final (called "poor riding qualities") were due to the vehicle's response to light turbulence.

After a series of reviews and briefings, Gentry was cleared to make a third glide flight. The initial attempt was made on August 8, 1969, but was aborted because of problems with a warning light, telemetry to ground control, and an uncomfortably warm cockpit. After landing Gentry showed signs of heat exhaustion and had a temperature of 103 degrees. The next attempt was rescheduled for August 21.

The countdown went as planned up to 45 seconds before launch. On this mission a new B-52 pilot was being checked out, and at this point he made a switching error—rather than arming the launch switch, he hit the launch switch. The X-24A was released. Gentry casually announced, "I've been inadvertently launched," and then flew the mission as planned. The cameras were not turned on, as that was done at 30 seconds before launch, but the photo chase coverage was good. As Gentry glided down, he performed a good pushover pull-up with the upper flap set at 15 degrees. Except for the test maneuvers, the flap setting was 21 degrees. The test showed no indication of flow separation at any time. By reducing the wedge angle of the flaps, the vehicle's lift-to-drag ratio was improved.

Thanks to the southerly drop point, Gentry could not reach Runway 18, the normal landing area. He switched to Runway 17 on the south lake bed and made a successful landing. Again, the landing rockets were fired at the flare, but the pilot was more comfortable with the approach this time, and he used the rocket for energy management rather than to stabilize the vehicle. The vehicle rolled out and came to a stop. Gentry then had to sit and wait a long time for the ground crew to arrive to get him out.

Between September 9, 1969, and February 24, 1970, an additional six glide flights were made to test various flap angles and rudder biases. All but one were flown by Gentry. The exception was the sixth glide flight, made by John Manke on October 22, 1969. The flights successfully qualified the lower wedge angle, which increased the lift-to-drag ratio. On January 23, 1970, a ground rocket test was done with the XLR-11 rocket engine, the first time the engine had been run in the vehicle. A month later, on February 20, an unmanned captive flight was done to test the propulsion system, pylon damping, and the checklist for powered flight. Following the ninth and final glide flight on February 24, 1970, the way was clear for the X-24A to make a powered flight.

CAPTAIN MIDNIGHT AND THE MIDNIGHT SKULKERS STRIKE

The NASA and Air Force personnel working on the lifting-body program had a good working relationship, without the corrosive rivalry of the X-airplane programs in the fifties. There did remain a low-level, good-natured rivalry, however. One expression of it was the markings on the different lifting bodies.

The X-15 had been a joint program of the Air Force and NASA and had carried "U.S. Air Force," "USAF," the national insignia, and "NASA." The X-24A was marked in the same way. The NASA-sponsored lifting bodies, the M-2s and the HL-10, only had "NASA" painted on their fins and the NASA meatball on the forward fuselages. Even though Air Force pilots, including Chuck Yeager, Jim Wood, Don Sorlie, and Jerry Gentry, had flown them, there was no sign of Air Force involvement. Gentry (a.k.a. Captain Midnight) decided to correct this.

Late one night (actually it was after everyone went home for the day) Captain Midnight and his fearless Midnight Skulkers sneaked into the NASA hangar where the lifting bodies were kept. They then proceeded to put "U.S. Air Force" on the NASA lifting bodies with stick-on lettering. They escaped undetected, and their NASA counterparts received a surprise the next morning.

Later, an "Official USAF Retouched Photo" was produced to commemorate the event. It showed Maj. Chuck Archie and Captain Midnight standing by the HL-10 parked on the lake bed. "USAF" had been retouched on the fuselage side, and a U.S. national insignia was added to the tip fin. Gentry was retouched with a mask and cape and the Captain Midnight insignia on his chest. The caption read, "The Midnight Skulker strikes again . . . or, It really is a joint lifting body program!"

Nowhere in the official histories of the lifting-body programs is this event recorded. It was, however, part of the spirit that made the lifting-body program successful.

THE X-24A POWERED FLIGHTS: THE UGLY DUCKLING SOARS

The X-24A's first powered flight was made on March 19, 1970. Gentry reached Mach 0.865 at 44,384 feet. Two chambers of the XLR-11 engine were fired for 160 seconds. The flight tested the rocket engine operation and the stability and control system, and it gathered data on vehicle handling during powered flight. The only problem was a badly worn left tire discovered after the landing rollout. This was followed on April 2 by the second powered flight, made by John Manke. He fired three of the rocket chambers, which pushed the vehicle to a slightly higher speed of Mach 0.866.

The X-24A then embarked on a series of flights to gather transonic stability and control, lift-to-drag, trim, and aerodynamic data at speeds above Mach 0.9. The first was on April 22 and was flown by Gentry. For the first time, all four chambers of the XLR-11 engine were fired, propelling the X-24A to Mach 0.925.

The next flight was made on May 14, 1970, and suffered from the first of a series of problems with the XLR-11 engine. Following the launch from the B-52, Manke began igniting the rocket chambers. Chambers 2 and 3 did not fire, and he was forced to fly an alternate profile. Rather than the planned speed of Mach 0.95, the actual speed was only Mach 0.748. Manke was able to record some data at speeds of Mach 0.7 and below. Coincidentally, it was the X-24A's thirteenth flight. There would be more engine problems in the future for the X-24A.

It took nearly a month before the problem was checked out and the vehicle was ready for its next flight. On June 17, Manke roared back in fine style, reaching a speed of Mach 0.99, only a few knots below supersonic speed. At Mach 0.9, the dampers were turned off. The pilot observed poor lateral control at a 5-degree angle of attack.

These powered flights had revealed an unexpected stability problem. While the rocket engine was firing, there were unexpected transonic pitch trim changes and a reduction in lateral directional stability. It was calculated that a 7-inch error between the engine centerline and the X-24A's center of gravity was required to explain the pitch trim changes. When the misalignment was precisely measured, it was found to be less than 2 inches. The engineers finally concluded that the pitch trim changes and the reduction in lateral directional stability were the result of the rocket exhaust plume affecting the airflow around the rear of the vehicle.

With the reduction in lateral directional stability, the planned low angles of attack could not be used for powered flight. It also resulted in a lowering of the vehicle's top speed. The wind tunnel data had predicted that the X-24A would have adequate stability at transonic and supersonic speeds with the upper flap held at a 35-degree angle. After several subsonic powered flights indicated that the engine exhaust was causing the problem, the upper flap angle had to be increased to 40 degrees to restore stability. The added drag meant that the projected maximum speed was reduced from Mach 1.8 to Mach 1.7.

The 40-degree wedge angle was tested on the two final flights in the test series, flown on Gentry's July 28 mission and by Manke on August 11. On the last flight, trouble with the engine again cropped up. Chamber 2 was 20 seconds slow in igniting. The problem did no harm, and a top speed of Mach 0.986 was reached. The new wedge angle was successful, and the way was now clear to expand the X-24A's envelope to supersonic speed.

PAINTING GENTRY'S CAR

Jerry Gentry drove an old 1954 Ford to work. It was in relatively good shape mechanically, but it had a few external bumps and bruises. Someone decided that it needed a new paint job. Preparations were secretly made to paint the car during Gentry's next lifting-body flight. The morning of the flight, Gentry drove over to Dryden and parked his car in his usual spot. He then walked to the van used to dress the pilots in their pressure suits. The minute the door closed on the van, the activity began. Project personnel fired up the car (Gentry normally left the key in the ignition) and drove it over to the paint shop. At the paint shop, they masked off the windows, bumpers, wheels, grille, headlights, taillights, and a few other pieces of trim, then began sanding the old paint to ensure that the new paint would stick. The new paint was not paint, however.

It was zinc-chromate primer, used on most aircraft parts to prevent corrosion. The bright yellow-green was visible for miles. The color resembled a fluorescent lime. It was obnoxious. Once the paint dried, the crew stripped the masking tape and paper off the car and began adorning it with large stick-on daisies approximately one foot in diameter (larger versions of the flowers on Bill Dana's space boots). The flowers came in a wide variety of colors that clashed with the car's paint scheme. To complete the decorations, flowers were stuck on the wheels and on several windows.

When the conspirators finally backed the car out into the sunlight, it was a dazzling sight to behold. They drove the car around the Dryden complex several times to allow everyone to observe their handiwork and then parked it back in its original parking place. When Gentry returned to Dryden after the flight, he noticed the car but did not realize that it was his. Only later, when he was preparing to drive home, did he recognize it. He was stunned. He had been involved in the flight for slightly over two hours, and during that short time his car had been grossly violated. The two-hour paint job was a tribute to the team of conspirators. You can do anything if you have enough eager volunteers. That group worked as efficiently as a race-car crew in the Indianapolis 500.

Gentry's sense of humor was strained somewhat, but he accepted the grotesque results and drove home to the delight of his neighbors and their children. They really made a fuss over that car. Gentry did not try to undo the decorations. He drove the decorated car for a couple of years before willing it to another test pilot when he left Edwards.

Gentry's adventure with his special car didn't stop with the paint job. One day the car wouldn't run. Gentry decided that he would tow it into Lancaster, a 30-mile trip, to have it fixed. The problem was that no one was available to help him. That didn't stop Gentry. He tied the inoperative car to his family car with a long rope and began towing it down the road. He was being cautious. He managed to tow the car off the base, about 20 miles, through the town of Rosamond, and within 5 miles of Lancaster before losing control. His inoperative decorated work car ended up in a ditch alongside the Sierra Highway. It's amazing that his family car wasn't dragged into the ditch also. Gentry didn't let little problems or obstacles stop him. He was a doer and a goer.

THE X-24A SUPERSONIC FLIGHTS

The first supersonic X-24A flight was scheduled for August 26, 1970, with Gentry as pilot. The flight was to expand the vehicle's envelope to Mach 1.1 and to acquire stability and control data. After launch from the B-52, Gentry attempted to fire the engine. His first two tries to ignite four chambers were unsuccessful. On the third try, two chambers fired, and he flew the alternate flight plan. The igniter failure was not the end of Gentry's problems. After landing, an inspection showed a fire had caused damage to the aft section during the jettisoning of fuel.

Instrumentation wiring and aluminum pressure-sensing lines were damaged by the fire, and one of the upper flaps was slightly warped by the heat. The telemetry signals from the engine showed it had performed normally during the burn, and the cause of the fire was unknown. The wiring and pressure lines were replaced, and the warped flap was replaced by a flap from an SV-5J in storage at Martin. The jettison lines were extended and canted outboard to prevent a recurrence. The engineers were also concerned that the fire might have weakened the metal in the engine mount. The engine and its mount were removed from the vehicle. John Cochrane, Martin's tech rep, and Norm DeMar, a NASA engineer, loaded the mount into Cochrane's Piper Comanche light aircraft and flew it to Martin's Denver facility, where it was annealed in heat-treatment ovens. The X-24A fire caused a six-week delay in the flight program.

The X-24A supersonic flight was rescheduled for October 14, 1970, with Manke as the pilot. It was an auspicious date; twenty-three years before, Chuck Yeager had made the first supersonic flight in the X-1. The X-24A's launch and engine ignition were successful, and Manke reached a top speed of Mach 1.186 at 67,900 feet. (Yeager had reached Mach 1.06 at 43,000 feet on the first supersonic flight.) In addition to expanding the vehicle's envelope to Mach 1.1, the flight also obtained aerodynamic data at supersonic speeds, tested the 40-degree upper flap setting, and made a 270-knot approach.

Manke's next flight, on October 27, was nearly as significant. The flight reached a speed of Mach 1.357 and an altitude of 71,407 feet, the record for the X-24A. The vehicle's descent simulated a space shuttle approach and landing. Manke touched down on lake bed Runway 18, making a precise, unpowered landing that added support to the idea that the space shuttle could land on a runway without engines.

On November 20, Gentry reached Mach 1.370 in what was the last X-24A flight of the year, and his last flight in the vehicle. Gentry had received an assignment to Vietnam, and to replace him, Maj. Cecil Powell had been selected. As with new HL-10 pilots, Powell made a glide flight on February 4, 1971, for pilot familiarization. It was followed on March 8 by Powell's first powered flight, a three-chamber transonic research flight to gather stability and control data. During the flight, Powell reached a speed of Mach 1.002 and landed successfully in a strong crosswind.

While Powell was entering the X-24A program, Manke was continuing the aerodynamic tests at supersonic speeds. On January 21, he made the first flight of 1971. The objectives of the flight were to expand the envelope to Mach 1.5, determine lateral directional derivatives, and gather longitudinal trim and lift-to-drag data with the 40-degree upper flap setting. After launch, however, the angle-of-attack indicator failed, and Manke had to shut down two of the rocket chambers and fly an alternate profile. Top speed was Mach 1.023.

Manke was more successful with a second attempt on February 18. The research goals were the same as the first attempt, and the pilot was able to make a number of good pushover pull-ups and aileron doublets at differing angles of attack. (For obvious reasons, it would not have been possible to fly those maneuvers on the previous flight with the broken angle-of-attack indicator.) Top speed was Mach 1.511 at 67,456 feet.

During the early powered flights, a pitch trim change was discovered at transonic speeds, which was caused by the engine exhaust plume. Manke again observed the change at speeds above Mach 1.3. The first time he encountered it, he thought the vehicle might diverge in sideslip (that it might try to turn sideways). The only thing that prevented him from shutting down the engine was that he still had excellent rudder control and so was able to hold the vehicle steady. The problem was that he had to spend time controlling the sideslip rather than flying the profile; an excessive amount of pilot effort had to be devoted to this for adequate performance. On the next flight, the angle of attack was reduced by only 2 degrees, which resulted in good handling. That underlined the unknowns of transonic flight; even small changes in Mach number and angle of attack can result in major aerodynamic changes.

Such problems required careful, painstaking analysis of the flight data, looking for subtle changes. On the February 18 flight, for example, the static pressure data from the flaps showed a significant increase in the base, or aft-surface, pressure with the engine running. There was also a further time correlation between the pitch trim change and a rather abrupt drop in the upper surface pressures to the same value as the under surface pressures, pointing to an interaction between the exhaust plume and the fuselage.

Despite the control problem, the envelope expansion had continued by using a reduced angle of attack. A flight control modification to solve the problem was introduced on the next flight. It consisted of a lateral-acceleration feedback loop, which allowed higher angles of attack to be used. The new control system was first tested by John Manke. He reached Mach 1.6 on March 29 in what was to be the fastest flight of the X-24A.

Subsequent X-24A flights would continue testing of the lateral-acceleration feedback loop and see the first use of the automatic control of the upper flap position. The automatic control had been developed by Martin and would change the position of the flaps according to the Mach number. It had not been used originally because of a lack of backups. Later in the program, it was found that the automatic system interfered with data collection and speed control during landing.

To acquire flight data, the test pilot had to fly a specific profile of speed, altitude, angle of attack, and control (i.e., flap) setting. To that extent, flight research was the same as any other scientific experiment conducted under controlled conditions. Rather than allowing the automatic system to control the flap settings, the lifting-body pilots would manually control them. As the vehicle went subsonic, the pilot would close the flaps, going from a high wedge angle to a low wedge angle in a single step. This was called a close-up or configuration change. During the landing, the pilots would briefly increase or decrease the vehicle drag by opening or closing the upper and lower flaps. The flaps acted like speed brakes.

The first flight test of the automatic flap control system was made by Powell on his May 12, 1971, flight. The peak speed was Mach 1.389, within the previous flight envelope. The way was now clear to push the X-24A to its maximum speed. One shadow over this effort was a reappearance of the XLR-11 engine problems seen on the earlier flights, in the form of a 20-second delay in engine ignition on Powell's flight. There was worse to come.

Following Powell's flight the X-24A was prepared for the maximum-speed flight. In addition to the simulator runs for Manke, the ground crew undertook more traditional preparations. The X-24A was given a complete polish job. About 100 pounds of equipment was also removed to lighten the vehicle. The maximum speed was expected to be about Mach 1.7.

Manke made the flight on May 25, 1971. After drop, the XLR-11 engine did not ignite properly, and only three rocket chambers fired. The pilot was forced to fly a three-chamber profile. The maximum speed was Mach 1.191. Despite the problems, all the subsonic data were collected, and the lateral-acceleration feedback loop was evaluated with power on. The feedback loop did not noticeably improve the vehicle's flight characteristics, so the gain would be increased on the next flight.

Because of the engine problems, a different XLR-11 was installed on the vehicle, and a 60-second ground run was made. The X-24A was weighed before and after the run in an attempt to measure the propellant flow rate. The preliminary analysis of the data indicated the flow rate with the new engine was higher but that it produced about the same amount of thrust as the old engine. This meant the engine burn time would be only 130 seconds, rather than the 141 seconds expected, resulting in a reduced maximum Mach number. The X-24A team reviewed the data and hoped they were wrong. The maximum-speed run, which was the twenty-eighth X-24A flight, was scheduled for Friday, June 4, with Manke as pilot.

The first sign of trouble occurred before launch. During the countdown, the chamber 3 rocket failed its prelaunch igniter test on the first try but was successful on the second attempt. With the engine now seeming to operate normally, the X-24A was launched.

When Manke tried to ignite the four rocket chambers, only chamber 2 ignited on the first try. The first attempt to ignite chambers 1 and 3 failed because of low liquid oxygen (LOX) manifold pressure, possibly due to LOX pump cavitation. Chamber 1 fired on the second attempt, but chamber 3 failed because the LOX valve didn't open. Chamber 4 never did fire, because of the failure of the pressure switch to activate and signal the propellant valves to open.

Manke flew the two-chamber flight profile with no further problems, but the research objectives could not be met with the engine problems. The Mach number was limited to 0.817, and the peak altitude was 54,373 feet. The landing was successful.

The postlanding inspection showed that the igniters in both chambers 3 and 4 had exploded, and there was evidence of an LOX leak in chamber 3 and a fuel leak in chamber 4. The XLR-11 engine had suffered multiple problems, several of which could not be explained.

Following the failed attempt, flight planning and simulator work began for another try at a maximum speed profile. Soon, second thoughts appeared. At best, a speed increase of Mach 0.1 could be achieved. When the value of a single added data point was balanced against the risk of losing the vehicle, it was decided to ground the vehicle, bringing an end to the X-24A program. Troubleshooting of the engine problems continued, including plans for a ground engine run, as well as writing reports, calibrations, and other activities.

In all, the X-24A had made twenty-eight flights, ten glide and eighteen powered. This was considerably less than the thirty-seven flights of the HL-10. The maximum speed reached by the X-24A was Mach 1.6, with a maximum altitude of 71,400 feet, both by John Manke. Both Manke and Cecil Powell subsequently joined the M2-F3 test program. Jerry Gentry would also have one last lifting-body flight before heading to Vietnam.

The X-24A, in terms of handling, was near the HL-10 and much better than the M2-F2. On its first two glide flights the X-24A showed a lateral sensitivity problem. Once the problem was attributed to erroneous wind tunnel data and minor changes in the control system were made, the vehicle flew well. The problems on the early glide flights were much less severe than the control problems shown by the other two original lifting bodies on their first flights. The later problems from the exhaust plume and the rolling moments caused by turbulence created some anxiety, but they were lessened with pilot experience.

One aspect of the landing in the X-24A was a strong nose-down pitch, which occurred as the gear was extended. The nose gear door was a flat plate that hung down at a 90-degree angle and disrupted the airflow on the underside of the vehicle. All three of the landing gear struts pivoted forward as they deployed. This created a forward movement of the center of gravity. The nose gear door was modified so it was at a 45-degree angle, resulting in a small reduction in the nose-down pitch. This pitching motion was undesirable because it occurred so close to touchdown and could easily interfere with the final approach and landing. (The increase in drag from the landing gear required the late deployment.) It caused apprehension among the pilots during the first few flights, but they soon learned to apply a little aft stick before gear deployment to counteract it. The landing gear also caused problems during rollout. Unlike the low-slung M2-F2 and F3 and HL-10, the X-24A had long landing gear struts, which raised the center of gravity relative to the width of the gear. The result was that the X-24A had poor crosswind landing characteristics. If the pilot tried to use differential braking and ailerons, the vehicle would lean to one side. If no correction was made, the X-24A would drift off the lake bed runway. Although the X-24A demonstrated the ability to make a precision landing following a simulated space shuttle descent, it could not land on the narrow confines of a concrete runway. (This would have been true even if the nosewheel steering had been activated, as the problem was due to the height of the gear.)

The X-24A was, to some extent, overshadowed by the other two lifting bodies. The X-24A program came after the M2-F2 and HL-10 programs had started, and the HL-10 flew faster, higher, and more often. The engine problems that aborted the X-24A's final mission and led to the program's end also cast a pall. One has the sense that the X-24A was cut short before it could show its full potential.

Despite the seemingly premature end of the X-24A, its story was not yet over. On December 15, 1971, a California Air National Guard C-130 transported the X-24A from Edwards to Martin's Denver facility. There, the ugly duckling would be transformed into a swan.

THE RETURN OF THE M2-F3

The "inspected" M2-F3 was returned to Dryden in July 1969, where nearly a year of additional work followed before the vehicle was considered ready for its first flight. Although the most visible change was the center fin, the original Northrop control system was replaced with one built by Sperry that was similar to that on the X-24A. Another addition was the reaction control augmentation system. It used several small thrusters, similar to the attitude control rockets used on spacecraft, rather than aerodynamic surfaces to provide stability control in the pitch axis. The rudders were also modified to allow them to act as speed brakes (something the original M2-F2 had lacked). Although the M2-F3 had the steepest descent angle, the pilots recognized the need for a speed brake to control the vehicle's energy on final approach. The rudders were simultaneously pivoted outward to increase drag, a method based on experience gained from the HL-10 and X-24A, both of which had speed brakes.

The M2-F2 accident had cost some $700,000 and just over three years of delay. What had been the first of the original heavyweight lifting bodies to make a glide flight was now the last to make a powered flight.

The project pilot for most of the M2-F3 flights was Bill Dana. He had mixed feelings about the vehicle. Years later, Dana said that he had wanted to fly the vehicle ever since seeing the M2-F1 and M2-F2 in flight, but he was also highly critical of the decision to keep flying the M2-F2 with the known PIO tendency. Dana developed an affection for the vehicle and the challenge it represented. He felt a sympathy for it, as one would toward a crippled child. Flying the HL-10 was not difficult; its handling qualities were so good that Dana felt anyone could fly it. The M-2 shape, however, because of its low lift-to-drag ratio, required the pilot's complete attention during landing. Any defects would make it unflyable. That was the challenge the M2-F3 represented.

The M2-F3 made its first glide flight since the accident on June 2, 1970. The flight was for pilot checkout and to evaluate the new center fin, to determine aileron characteristics, and to evaluate the modified stability augmentation system and speed brakes. Despite all the wind tunnel tests done since the accident, there was still apprehension about the vehicle. Jack Kolf told Dana to look at the attitude indicator after launch. Commonly called the eight-ball, this instrument showed the vehicle attitude and was labeled "N," "S," "E," and "W" for the cardinal directions. Kolf said, "If the heading reads M, eject."

The transformation of the M2-F3 was immediately apparent after launch. The vehicle was rock steady in roll. A graph of the M2-F3 stick movements was smooth. In contrast, a similar graph of an M2-F2 flight was irregular, as the pilot had to move the stick constantly to keep the vehicle stable. The difference between the two vehicles was apparent from a close examination of landing films. The M2-F2 could be seen to twitch as the pilot had to make corrections. In contrast, the M2-F3 held steady throughout the landing approach. Only two more glide flights, on July 21 and November 2, 1970, were required before the vehicle was cleared for a powered flight.

TOWARD MACH 1

The M2-F3's first powered flight was made on November 25, 1970. The goal was to expand the envelope to Mach 0.8, check out engine operation, obtain stability and control and other aerodynamic data, and check landing visibility. The vehicle reached a speed of Mach 0.809, but it had stabilized at a higher angle of attack than expected. As a precaution, Dana shut down the engine and landed on a different runway from the one planned.

Dana felt confident that he could land the vehicle without using the nose windows. As a test, they were covered with paper on the first powered flight. A cord would allow Dana to rip them off if the landing proved too difficult. He made a successful but hard landing and concluded that the experiment should not be tried again.

On February 9, 1971, Jerry Gentry made a glide in the M2-F3. He was the last of the M2-F2 pilots still active, and the flight was planned to compare the handling characteristics of the M2-F3 and the original M2-F2. After landing he said that it handled as well as the HL-10; this was high praise given the HL-10's reputation and Gentry's dislike of the original M2-F2. Years later, he said that the center fin had "solved the problem of the adverse yaw and made it fly very pleasantly." It was Gentry's last flight before leaving for Vietnam. He had the distinction of being the only pilot to fly the M2-F1, M2-F2, HL-10, X-24A, and M2-F3. February 9 was marked not only by Gentry's glide flight but also by the Sylmar earthquake, the Apollo 14 splashdown, and a lunar eclipse.

Two weeks later, on February 26, Dana made the next M2-F3 powered flight. After launch, the first of several problems occurred. When Dana fired the engine, only two of the chambers lit. The pilot switched to a cruise-back flight plan, and most of the objectives were met. Top speed was Mach 0.773 rather than the Mach 0.85 planned. After engine shutdown, Dana jettisoned the remaining propellant in order to lighten the vehicle as quickly as possible. As this was being done, a small fire broke out in the tail, which continued to burn until the jettisoning was complete. Finally, as the vehicle was on final approach, Dana pulled the gear release handle and nothing happened. Apparently some water had frozen and the ice had jammed the gear release. Dana had to pull extremely hard before the gear came down. All the while, the lake bed was rushing up toward the vehicle.

The M2-F3 was grounded for several months to study the problems. In mid-

May, a ground engine test was planned to check out the repairs. During preparation, a ground crewman noticed that fuel, a mixture of alcohol and water, was leaking out of the liquid oxygen vent. Because of multiple failures in one of the servicing lines, fuel had flowed into the LOX tank. The resulting mix of LOX, alcohol, and water was as explosive as nitroglycerin. The ground crew opened the LOX tank vent valves and allowed the LOX to boil off. The area was evacuated, and all supersonic flying in the Edwards area was canceled to avoid the real possibility that a sonic boom might cause an explosion.

It took several days before the LOX had completely boiled off. The tank was purged of fuel, the faulty valve was replaced, and a successful ground run was completed on May 20. The next step was a captive flight under the B-52 to check out the modifications to the landing gear deployment system. The M2-F3 captive flight was delayed twice because of wind and clouds at Edwards.

The captive flight was finally made on June 1 with poor results. When Dana pulled the landing gear handle, the left main gear deployed about a second behind the other two. The cause was unknown, but it could have been airloads from the B-52 or a frozen door or uplock, since the LOX vent had been repositioned directly in front of the left main gear. More instrumentation was installed on the gear, and another captive flight was planned. With the time required to install the instrumentation and for the B-52 to complete its phase inspection, it was not until July 1 that the captive flight was successfully made.

The first envelope expansion flight was made by Dana on July 23. The vehicle reached a speed of Mach 0.930 to obtain stability and control data at Mach 0.8 and to evaluate the reaction control system. This was followed by another flight on August 9, which reached Mach 0.974. On this flight the launch altitude was reduced to 40,000 feet from the usual 45,000 feet to decrease the launch Mach number.

The M2-F3's first supersonic flight was made on August 25, 1971. Maximum speed was Mach 1.095, and a peak altitude of 67,300 feet was reached. In addition to expanding the speed envelope, the flight also obtained data on aileron effectiveness at speeds greater than Mach 0.9, along with stability and control data at Mach 0.95 and 0.9, and evaluated the speed brake. The M2-F3 had taken just five powered flights to reach supersonic speeds. The HL-10 had also taken five powered flights to reach Mach 1, and the X-24A, held up by engine problems, had required nine. Both the HL-10 and X-24A programs had been completed, with the final X-24A flight having been made nearly three months before.

XLR-11 ENGINE PROBLEMS

Like the X-24A before it, the M2-F3 also went through problems with its XLR-11 rocket engine. Following the first supersonic flight, the next mission was scheduled for September 24 to evaluate stability and control at Mach 1 and test the reaction control system. An ignition malfunction in two engine chambers required the other chambers to be shut down prematurely. Dana jettisoned the propellant, sparking a small fire in the tail area. The jettison was terminated, extinguishing the fire. Dana made a heavyweight landing on the Rosamond lake bed.

This was the second fire on the M2-F3, and it followed a similar fire on an X-24A flight. Engineers were finally able to isolate the cause. Following engine shutdown, the remaining fuel in the engine chambers continued to burn (called an after-fire). When the LOX and fuel were jettisoned, they mixed in the turbulent airflow between the upper and lower flaps and were then ignited by the after-fire. They continued to burn as long as the propellants were available. This combustion had not been seen on the other flights because the normal jettison altitude was so high that there was not enough atmospheric oxygen to sustain the after-fire. (Also, alcohol burns with a clear flame.) As a result, the jettison system was modified and an engine chamber purge system was added to prevent the after-fires. Dana made a glide flight on November 15 to test the new systems. It was successful, and the research flights could now resume.

The next powered flight was made on December 1, 1971. Dana reached Mach 1.274 and 70,800 feet. The objectives—flight envelope expansion, stability and control data above Mach 1, and evaluation of the reaction control system—were successfully accomplished. The vehicle was quickly turned around and scheduled for a flight on December 16 with the same objectives as the previous flight.

After launch, Dana started the chambers individually. A pump overspeed and low LOX manifold pressure shutdown occurred as the number 2 chamber was trying to stabilize. Dana flew a three-chamber profile; with the engine problem, the only objective that could be achieved was the reaction control test. As the vehicle descended, NASA pilot Tom McMurtry was flying chase in an F-104. As he tried to follow the M2-F3, he put in too much rudder, and at 37,000 feet, the F-104 went into a deep stall and flat spin. Fortunately, the plane was rocking in pitch. The amplitude of the pitching motion increased, and with full forward stick, McMurtry broke the spin. He pulled out slowly to avoid a secondary stall, finally recovered at about 18,000 feet, and rejoined Dana on the landing approach. Dana's landing was uneventful, but at the debriefing all the questions were about McMurtry's spin. The Advanced Vehicle Section's weekly report noted, "He stole the show."

COMMAND AUGMENTATION SYSTEM TESTS

The lifting-body program went on hiatus for the first half of 1972. There were several reasons. The B-52 launch aircraft was scheduled to leave Edwards on December 29 for a major inspection and overhaul, which would take several months. The M2-F3 was itself grounded in January 1972 for installation of the command augmentation system (CAS). The CAS was the first fly-by-wire control system developed for a lifting body. It included a pitch-rate command system, which allowed the pilot to set exactly the amount of pitch change desired. The other feature was an angle-of-attack hold. Once the pilot had established the desired angle of attack, the hold would be engaged and the system would automatically maintain it. The CAS was to improve the vehicle's handling and reduce the pilot's workload.

It was not until July 1972 that the M2-F3 was ready to fly again. After two launch attempts were aborted, the first CAS flight was made on July 25. It was subsonic, with a maximum speed of Mach 0.989 and a peak altitude of 60,900 feet. This was followed by a series of flights to increasing speeds. On August 11 the M2-F3 made its third supersonic flight, reaching Mach 1.101, and on August 24, it reached Mach 1.266. The flight of September 12, which came only a few days after the celebration of Dryden's twenty-fifth anniversary, saw a reoccurrence of the engine problems. Two of the four chambers failed to ignite, limiting the speed to Mach 0.880 and preventing any of the test objectives from being fulfilled. The program quickly regrouped, and on September 27 Dana reached a speed of Mach 1.340. On October 5, the maximum speed was increased to Mach 1.370. This flight, the nineteenth of the M2-F3 program, was a milestone for another reason. It was the one-hundredth flight of a lifting-body vehicle.

Up to this point, Bill Dana had flown all the M2-F3 flights except Jerry Gentry's single glide flight. During October and November 1972, two more pilots, John Manke and Cecil Powell, joined the program. Unlike the drill for the X-24A program, their first flights were subsonic powered flights rather than glide flights. Manke's first M2-F3 flight was on October 19; Powell's was on

November 9. They were followed by a supersonic familiarization flight for each (Manke's on November 1 and Powell's on November 29).

The speed buildup continued as the year, and the program, neared its end. On November 21, Manke reached Mach 1.435 in an evaluation of the CAS, the angle-of-attack hold, and the pitch reaction augmentation system. Fall rains forced a change in plans. The Rogers lake bed was wet, so the landing site was switched to Rosamond. Runways 02 and 20 were selected for both normal and emergency landings.

During the boost phase of the flight, Manke observed "lateral-directional upsets," which had appeared during his HL-10 and X-24A flights. They had originally raised considerable pilot concern that the vehicle might uncork. They were unlike the behavior of conventional fighter aircraft, and this raised a red flag until it was understood. As the pilots gained experience, they realized that the upsets were not serious. It had long been suspected that they were caused by horizontal windshears, but proof had been lacking. On this flight, a movie camera had been set up to photograph the breakup of the M2-F3 contrail due to winds. Reasonable correlations were found between two large windshears shown on the film and two lateral disturbances in the data.

The CAS continued to be tested on the final three flights of the program in December. Cecil Powell made his final lifting-body flight on December 6, reaching a speed of Mach 1.191. The flight also provided stability and control data at Mach 1.1, 0.8, and 0.7 and checked out the pitch reaction control. He successfully landed on the Rosamond lake bed. The final two flights, like those of the other lifting bodies, were planned to reach the vehicle's speed and altitude limits.

Dana made the maximum-speed flight on December 13, 1972. To obtain data on stability and control at the vehicle's maximum Mach number, several procedures were used. The launch from the B-52 was made at 47,000 feet rather than the usual 45,000 feet. As the XLR-11 engine burn neared its end, Dana fired the landing rockets to gain the last possible bit of speed. The vehicle reached a maximum Mach number of 1.613. The M2-F3 was quickly turned around, and on December 21, Manke made the maximum-altitude flight, reaching 71,500 feet. Much of the flight was made with the aerodynamic pitch damper off, with control being maintained using the reaction control system.

In all, the M2-F3 had made twenty-seven flights in two and a half years. Of these, fourteen had been made in 1972, with half flown in November and December. (Another sixteen glide flights had been performed as the M2-F2.) The

M-2 shape was both the first and the last of the original lifting-body shapes to fly. John Manke later made extensive comments on the M2-F3's handling and how it compared with the other lifting bodies.

Manke found that the most difficult aspect of flying the M2-F3 was the boost phase following launch, as it was impossible to control the angle of attack precisely, because of trim changes at transonic speeds and shortcomings in the elevator system. After engine shutdown, the lateral stability of the M2-F3 was outstanding, except when the stability augmentation system was off and at low angles of attack. Under those conditions, roll sensitivity was high, much like the X-24A. The rudder sensitivity was low, but it was used only for test pulses and was not a problem. (The rudder-to-aileron interconnect was removed from the M2-F3 after the first few flights.)

The M2-F3, like the HL-10 and X-24A, showed excellent handling qualities during approach and landing. With its lower lift-to-drag ratio, the flare had to be commenced at a higher altitude, but this was not a problem. The M2-F3 had better "riding qualities" in turbulence than either the HL-10 or X-24A. Its response to turbulence was not as quick as the other two vehicles', and the response was in the form of acceleration rather than a roll moment. Unlike the HL-10, the forward window did not distort the landing view, although depth perception was not good during the last 5 feet until touchdown. Rollout was similar to the X-24A, although with less directional control.

Manke found that the combination of the CAS and angle-of-attack hold "provided a near ideal control system." The pitch-rate command loop could hold the vehicle precisely at a given pitch angle, important given the poor manual pitch control during the boost phase. With the angle-of-attack hold engaged and the side stick controller centered, the pilot knew exactly what the vehicle would do. This reduced the pilot's workload, allowing him to concentrate on housekeeping chores, such as preparing for landing.

Manke considered the reaction augmentation system (RAS) for the pitch axis one of "the most interesting aspects of my exposure to the M2-F3." The pitch axis was considered critical; it was universally believed that a failure of the pitch stability augmentation system would make it impossible to land the vehicle. On Manke's last flight, during which he reached the M2-F3's maximum altitude, the RAS provided the only pitch damping used during several critical portions of the flight. At worst, Manke considered it half as effective as the normal system, but in other areas he was hard put to tell any difference. Manke said later that he had no qualms about landing the M2-F3 using only the RAS.

THE X-24B: FROM UGLY DUCKLING TO SWAN

While the M-2, HL-10, and SV-5 shapes were being developed in the sixties, the Air Force Flight Dynamics Laboratory was also looking at exotic vehicles. Unlike the original lifting bodies, these shapes had hypersonic lift-to-drag ratios of 2.5 to 1 or better. In contrast, the ratios of the original lifting bodies were between 1.2 and 1.4 to 1 at hypersonic speeds. The shapes, designated FDL-5, 6, and 7, were primarily intended as aircraft capable of prolonged flight at Mach 8 to 12 using air-breathing engines. The shapes were triangular, with stub wings, flat undersides, and rounded tops. Because some of the shapes had poor subsonic lift-to-drag ratios, they had flipout wings to add lift during landing.

Although not primarily intended as such, the FDL shapes also had applications as lifting bodies. The original lifting bodies could choose a landing site within an area the size of the United States, but the FDL shapes could touch down in an area stretching from Alaska to Greenland and southward to the upper part of South America. The shapes, with a larger lift area and the stub wings, also offered much better handling.

As with the first-generation lifting bodies, the FDL's subsonic and transonic handling needed to be tested at low cost. The original idea of Al Draper and Bill Zima was to build a gloved external shape that could be fitted around one of the Martin SV-5J jet-powered trainers. The new configuration was named the FDL-8.

In January 1969, the Flight Dynamics Laboratory issued a proposed development plan for just such an aircraft that would be launched from the B-52 (thus avoiding the problems of a ground takeoff). Like the X-24A, this was to be an Air Force–sponsored project. As the studies progressed, the advantages of rocket propulsion, such as speed and altitude, became clear. The Air Force scrapped the plan to use an SV-5J and instead began to fit the FDL-8 shape around the X-24A. Rather than $5 million for a new vehicle, converting the X-24A would cost only $1.1 million.

As with the original X-24A, the FDL-8X (as it was first known) soon became a joint Air Force and NASA program. In August 1970, the Air Force Flight Test Center commander and Paul Bikle endorsed the program. Air Force Systems Command delayed approval, however, pending arrangements for the joint funding. NASA program manager John McTigue forced the issue on March 11, 1971, by forwarding $550,000 directly to the Air Force Flight Dynamics Laboratory. Six weeks later, the Systems Command's director of laboratories gave his approval, and the program was under way.

On June 4, 1971, the X-24A made its final flight. Preparations then began for what was now called the X-24B. The subsystems were put in storage, and the shell was sent to Martin's Denver facility in December 1971. Several of the people who had built the original X-24A worked on the conversion, including Dick Boss, the chief of manufacturing. The X-24B team was also assembled: Jack Kolf of NASA was named the Dryden program manager, while Johnny Armstrong and Norm DeMar remained the Air Force program manager and NASA operations engineer. The envelope expansion and flight planning were under the direction of Robert Hoey, of the Advanced Vehicle Section.

The original prime Air Force program pilot was Maj. Cecil Powell, with Maj. Mike Love as his backup. Powell had been at Edwards for a considerable time by then, and he was transferred before he had a chance to fly the X-24B. Mike Love became the primary Air Force pilot, and John Manke served as NASA pilot.

Over the next ten months, the X-24A was transformed. The potatolike fuselage was hidden by a new triangular outer skin. The change in the vehicle's center of gravity and the added weight required a new nose gear. A Navy F11F nose gear, which had a working steering system, was selected. Other equipment came from the T-38, F-104, B-57, F-106, and X-15. On October 24, 1972, the completed X-24B was loaded on a C-5 transport and flown to Edwards. The ugly duckling was now a swan.

FROM GROUND TESTS TO POWERED FLIGHTS

After the X-24B's arrival, Norman DeMar began overseeing the reinstallation of the subsystems. The left side of the vehicle and the left tip fin were fitted with research instrumentation for correlation with the wind tunnel measurements. There were problems and delays. In December 1972, the cockpit electrical rework was running behind schedule, and it would be another month before power could be applied. By the summer of 1973, the work was complete, and the X-24B was ready to begin its initial tests.

Since December 1971, the Air Force X-24B team had been thinking about nose steering and ground stability in crosswinds. The Flight Dynamics Laboratory was also worried about nose gear shimmy and the heavy loads that would be placed on the gear. As with the X-24A, the first tests of the X-24B were ground taxi tests, referred to as the Bonneville Racer Runs. The first, a series of six low-speed runs with the landing rockets, were performed on July 2, 1973. On July 5 and 6, high-speed runs were made using two chambers of the XLR-11 engine. After an unmanned mated taxi test and a captive flight with John Manke in the cockpit for systems checks, the X-24B was ready for its first glide flight.

The first attempt at a glide flight was aborted after a gyro failure in the X-24B on July 24, 1973. It was rescheduled and successfully accomplished on August 1. The launch was made from the B-52 at an altitude of 40,000 feet. Manke maneuvered the vehicle to evaluate its handling and made a practice flare at 30,000 feet. During the flight, he had to fly through a small cumulus cloud that had formed along the planned flight path. It caused no problems, and touchdown on Runway 18 came 4 minutes and 11 seconds after launch. This was followed by a second glide flight on August 17, which investigated the transonic configuration, longitudinal trim, stability and control, fin and rudder pressure, and landing gear dynamics. Part of the flight was made with the dampers off.

The engine fires on the X-24A and M2-F3 had caused the Flight Dynamics Laboratory to perform a flow visualization wind tunnel test using a model of the X-24B. The test was looking for a fuel jettison location that would prevent fuel from recirculating into the rear fuselage, where it could be ignited by an engine after-fire. The location selected was at the side of the right fin, just above the aileron. The third glide flight, on August 31, was to test the jettison location. The vehicle was serviced with a small amount of the alcohol-water fuel. When Manke operated the fuel jettison, the X-24B rolled to the right, requiring 70 percent of the available aileron to bring the vehicle back to level flight. When the jettison ended, the X-24B resumed its normal behavior. Apparently the high-pressure stream of fuel had disrupted the airflow over the fin and aileron, causing the uncontrolled roll. The line was relocated to the rear of the fin, so the fuel was jettisoned directly aft. The new location was successfully tested on the fourth glide flight, made on September 18. At this point Mike Love joined the X-24B program, making a captive flight on October 3 and his first glide flight the following day.

After five glide flights, it was now time to begin powered flights. This did not prove easy. Manke's first attempt was aborted on October 31 when the engine igniters failed a prelaunch check. Another abort, on November 13, was due to clouds over the launch and landing sites. The skies had been clear when the

B-52 took off, but conditions had deteriorated during the flight to the launch point.

Success finally came on November 15. Manke was launched from the B-52, ignited three rocket chambers, and reached a speed of Mach 0.917 to obtain stability and control data. A month later, on December 12, he increased his speed to Mach 0.993. The objectives of the flight were also increased: stability and control at Mach 0.9, aileron bias at 11 degrees, fin and rudder pressure survey, and acoustic noise and vibration data.

SUPERSONIC FLIGHTS AND ENVELOPE EXPANSION

Flights did not resume until February 15, 1974, when Mike Love made a second glide flight for pilot training. This was followed on March 5 by the first supersonic X-24B flight, when Manke reached a speed of Mach 1.086. This was only the third powered flight by the vehicle. The summer and fall would be spent in an envelope expansion program. Following Love's first powered flight on April 30, the effort began in earnest.

The added weight of the X-24B posed a major performance problem. Simulator tests had shown that with the XLR-11 engine the vehicle could reach a speed of only about Mach 1.4. Two propulsion engineers, Jerry Brandt from the Air Force and Bill Arnold of Reaction Motors, the engine manufacturer, proposed increasing the chamber pressure to 300 psi, which would increase thrust from 8,600 pounds to 9,800 pounds. Ground tests showed this was safe, and the engine was modified. The engine would be started at the normal chamber pressure of 265 psi. Once all the chambers were running smoothly, which took about 30 seconds, the overdrive switch would be thrown by the pilot, increasing the chamber pressure regulator setting. The switch was added after Love's first powered flight.

On May 24, Manke reached Mach 1.14 in the first use of the overdrive switch. The intended speed was Mach 1.25, but one chamber failed to light, and the three-chamber profile had to be flown. The rest of the test objectives were met. Love raised the speed to Mach 1.54 on August 8. This flight also had problems; the engine suffered from pogo, an oscillation so named because it resembled the motion of a child on a pogo stick. Manke's August 29 flight was intended to reach Mach 1.68, but the engine shut down early, limiting the speed to Mach 1.097. The cause was a split in the fuel tank, which grounded the vehicle for two months.

The X-24B did not return to the skies of Edwards until October 25. When it did, Love reached the aircraft's maximum speed, Mach 1.752, second only to the HL-10's top speed as the fastest a lifting body had flown. Two more flights were made before the end of the year. Manke's November 15 flight reached Mach 1.615, but the LOX pump cavitated, and the pilot had to restart chambers 1 and 3. Love reached Mach 1.585 on December 17, the seventy-first anniversary of the Wright brothers' first flight at Kitty Hawk.

The final two flights of 1974 ushered in the third phase of the X-24B program, the research flights. They were followed by another eight flights between January and June 1975. The objectives were similar for each flight: stability and control data at different speeds and aileron biases; tests of different flap settings both in flight and during the landing approach; surveys of the fin, rudder, and flap pressures; and qualification of the space shuttle thermal protective system, or heat-shield tiles. Other tests undertaken on individual flights included boundary-layer noise and vibration experiments, checks of handling qualities with all dampers at zero gain and other control-system tests, and measures of the effects of the rocket plume on directional stability. John Manke and Mike Love, now a lieutenant colonel, took turns flying the missions. Speeds were between Mach 1.4 and 1.6, and the maximum altitude for the program was set on Manke's May 12, 1975, flight at 74,130 feet.

X-24B RUNWAY LANDINGS AND PILOT CHECKOUT FLIGHTS

The aerodynamic testing was concluded with Love's July 15, 1975, flight. With previous lifting bodies, the final flights had been maximum speed and altitude missions. With the X-24B, something completely different was planned. The FDL-8 shape was intended to be a hypersonic aircraft design. From the start, the X-24B team had wanted to attempt a runway landing with the vehicle. The nosewheel steering, braking, and cockpit visibility had all been designed with this in mind. The vehicle's ability to make a precise landing was tested on flights 16 through 24. (The first, on October 25, 1974, was the maximum-speed flight, and the last was made on June 6, 1975.) The pilots tried to land on a marker painted on the lake bed runway. The nine flights showed the X-24B pilot could land within 500 feet of the mark. With the X-24B moving at 320 feet per second at touchdown, that meant the pilot was within the landing zone for less than 3 seconds.

Based on those flights, Manke and Love concluded that the X-24B's handling was good enough to attempt a landing on the 15,000-foot concrete main runway at Edwards—something that had never been attempted with a rocket-powered research aircraft. Their proposal was submitted to the Flight Dynamics Laboratory, to Dryden, and to the Air Force Flight Test Center commander, Gen. Robert Rushworth, a former X-15 pilot. A runway landing by the X-24B would also be an important demonstration for the space shuttle.

Approval was given, and Manke, Love, and the program team began an intensive planning and training effort. Manke and Love made a three-week series of T-38 and F-104 training flights to simulate the X-24B's approach and landing behavior. Manke flew more than a hundred simulated approaches.

Manke made the first runway landing flight on August 5, 1975. After launch, he ignited the engine, but one of the four chambers failed to ignite, yet another of the engine problems that had beset the program. The pilot flew the three-chamber profile, reaching Mach 1.19 and 57,050 feet altitude. He aimed for a stripe painted on the runway 5,000 feet from the approach end. One of the main gear touched down 5 feet before the stripe, and the other touched down just after passing over it. Love made the second runway landing on August 20, 1975. After reaching Mach 1.548 and 71,040 feet, he landed 400 feet past the runway stripe.

Both pilots found that a runway landing was, in some ways, easier than a lake bed landing. The runway markers, roads, Joshua trees, and other visual clues gave additional "how goes it" information not available on the featureless lake bed. One major difference was the landing roll. The concrete runway was much smoother than the lake bed, so the rollout was longer. The runway landings used a different approach procedure from the standard on earlier lifting-body landings. At 1,000 feet, the X-24B would make a preflare, in which a shallow glide slope (about 3 degrees) would be established. Just before landing, a final short flare would be made. This two-step flare would later be used for space shuttle landings.

The final two powered X-24B flights were made by Bill Dana. The first, on September 8, 1975, reached a speed of Mach 1.48. His second X-24B flight, on September 23, was also the last powered flight of a rocket research aircraft. The era that had begun with the first glide flights of the X-1 was ending. For the occasion, Dana again wore the pink boots. After launch, one of the chambers failed to fire, and yet again the vehicle flew a three-chamber profile. The landing was normal, and the X-24B rolled to a stop on the lake bed. The four chase planes, two T-38s and two F-104s, formed a diamond formation and roared low overhead.

That night, the X-24B team held an end-of-an-era party at the Longhorn near

Lancaster. The mood was different from the many other postflight parties held over the previous three decades. It was a wake, rather than a party, as more than an era was ending. In 1966 a photo was taken of the main hangar at Dryden. Visible were the HL-10, M2-F2, M2-F1, an F-4, the F5D and F-104 chase planes, the C-47 tow plane, and the three X-15s. Now, with the end of the X-24B program, all that remained were the F-104s. Everyone knew those days were gone and would never return. Dryden would now undertake boring aerodynamic research, rather than the cutting-edge activities that had been its hallmark for the previous three decades.

There was still one final series of X-24B flights left to be made. With the powered research flight over, the vehicle was to be used for pilot familiarization glide flights. Three pilots were selected: Einar Enevoldson and Tom McMurtry of Dryden and Air Force captain Francis R. Scobee. Enevoldson made the first glide flight on October 9, 1975, followed by Scobee and McMurtry on October 21 and November 3. Each made a second glide flight in November, with McMurtry making the vehicle's thirty-sixth and final flight on November 26.

The X-24B was a second-generation lifting-body shape, and it benefited from a decade of experience. Like the other lifting bodies, it was affected by turbulence. Thanks to its triangular shape and stub wings, however, the effect was less than on the earlier vehicles. The X-24B had a lower dihedral effect than many conventional aircraft. The vehicle's handling was superb; Bill Dana thought it flew better than the HL-10, and John Manke observed that he had made numerous X-24B flights with the dampers off and noticed little difference in handling. In fact, he thought that if a new pilot flew the X-24B with the dampers off, he would not realize it needed them. One handling problem the X-24A and B both shared was trim changes due to the exhaust plume.

Like the X-24A and M2-F3, the X-24B had also experienced a long series of XLR-11 engine problems. The cause was the technological age of the engines. They had been designed in the mid-forties and lacked many of the features of later engines. Added stress was placed on the engines by the upgrades, particularly the increase in thrust.

The original three XLR-11s used in the lifting-body programs were the same engines used in the X-1 aircraft between the late forties and the late fifties. When the M2-F2 program began, there were no XLR-11 engines available at Dryden. Program director John McTigue actually had to take them from museums. One engine had come from the X-1 in the Smithsonian that Chuck Yeager had flown in the first supersonic flight, another came from the X-1B in the Air Force Museum, and the third came from Dryden's front yard—the

X-1E on display in front of the headquarters building. Spare parts had been found in Dryden warehouses or were built from scratch, while the engines were reconditioned by NASA mechanics who had worked on the original rocket-plane programs. Those three engines were still being used on the X-24B a decade later.

Throughout this period, there had been an ongoing debate about what form the next step would take. It centered on several issues. Would the next-generation spacecraft be a large capsule, or would a lifting-entry design be used? If it was to be a lifting-entry design, would it be a lifting-body shape or a winged spacecraft? The most contentious issue of all was the question of how a lifting-entry spacecraft would land: would it glide to a landing, or should it be powered?

As the X-24B team was mourning the passage of one era, another was about to begin. Across the desert, space shuttle *Enterprise* was taking shape.

10
THE LIFTING BODIES AND THE BIRTH OF THE SPACE SHUTTLE

By 1969, I had come to several personal opinions about future spacecraft designs. First, I definitely believed that a lifting-entry, horizontal landing, manned spacecraft was practical. I thought we could design a vehicle that is stable and controllable from orbital speeds down to landing speeds. I based this on the results of the X-23A vehicle tests, the X-24A flight tests, and the HL-10 flight tests.

Second, I believed that a pilot could make successful entries and accurately maneuver a lifting-entry vehicle from orbital altitude and velocities down to a precise landing on a preselected runway. This was based on my own experience flying entries in the Dyna-Soar simulator and flying actual entries from over 200,000 feet in the X-15. With no onboard propulsion and only the crudest form of energy management (ground controller callouts) and guidance (compass) assistance, I had maneuvered the X-15 from hypersonic speeds down to precise landings on a lake bed runway. I had also landed the M2-F1 and M2-F2 on the same runway. I concluded that we already had the technology and experience needed to build such a vehicle, that it was entirely practical, and that we should get on with it.

Since the program began in the early sixties, its ultimate goal had been to put a manned lifting body into orbit and to make a lifting reentry. The lifting-body concept had many critics, and that would be the final proof of its validity as a spacecraft design. It was a long-standing tradition of NASA research aircraft planning (going back to the NACA days) that even before a new aircraft made its first flight, its successor was under study. The same was true for the lifting bodies. Even before the M2-F2 made its first glide flight, we were thinking about putting one in orbit.

THE ORBITAL M2-F2 PROPOSAL

Hubert M. "Jake" Drake was conceptualizing this next step. In 1964 he began putting together a proposal to orbit the M2-F2 using a Titan 2 booster that had been modified to launch the Dyna-Soar on a suborbital trajectory. The modified Titan 2 booster never materialized because the Dyna-Soar was switched to a Titan 3 booster for orbital flight. The modified booster was, however, subsequently used to orbit the Gemini capsules.

The Gemini use resulted from a conversation between Walt Williams and me. Williams was the deputy director of the Space Task Group for Operations at that time, and he was searching for a booster for the Gemini capsules. I suggested the modified Titan 2 that had been proposed for Dyna-Soar because it had been beefed up structurally to withstand the aerodynamic loads imposed on the booster by the Dyna-Soar, which acted as a canard on the booster.

Jake Drake selected the same booster for the M2-F2 for many of the same reasons. The Titan 2 was capable of orbiting a payload of 7,500 pounds. The M2-F2 weighed less than 6,000 pounds without any propellants. If the orbital weight of the M2-F2 could be constrained to less than 7,500 pounds, the Titan 2 should work.

Drake asked me to work with him on the conceptual design of the orbital M2-F2. I was flattered. He had been a visionary and innovator long before I began work at the NACA, and he had been instrumental in the advocacy and generic design of the X-15. Drake was the person who selected the B-52 as the X-15 mothership, after evaluating a series of candidates including the B-36 and the B-58. He subsequently proposed air-launching a 1,000-pound payload into orbit from the B-52 using a large solid rocket booster. Submitted shortly after *Sputnik* was launched, during the frantic search for a means of launching a U.S. payload into orbit, the proposal was considered too radical and was not approved. Thirty years later we were launching a Pegasus booster and payload from a B-52 into Earth orbit.

I eagerly agreed to help Drake in any way I could. I was given the task of selecting the various subsystem components after conceptually defining the systems. We planned to use existing systems and components wherever possible.

Although the task was similar to what I had done in the development of the M2-F2, it was a real challenge to identify the necessary components and yet maintain an absolute launch weight limit.

The plan was to use the basic M2-F2 for the orbital mission with some major systems modifications to extend the operational flight duration and to cope with exoatmospheric and entry flight conditions. We added reaction controls for exoatmospheric flight, for example, using X-15 reaction control rockets because they were available and would do the job. We also planned to use the X-15 auxiliary power units with larger peroxide tanks to increase the operating time to two hours for an orbital mission. We proposed an ablative heat shield to protect the aluminum skin of the M-2 during entry. The pilot's canopy would be modified to look like the X-15 canopy, with small high-temperature glass windows. An inertial platform and computer would be installed for guidance and navigation information.

Drake and Don Bellman did the performance studies and the stability and control analysis to ensure we ended up with a realistic proposal. This conceptual design involved several months of effort, but we finally defined an M-2 vehicle that we believed would accomplish an orbital mission and safely survive reentry heating. The draft copy was finished on September 14, 1964.

We planned one unmanned suborbital flight to test the booster and heat shield at orbital speeds. The flight would be relatively short, ending in the mid-Atlantic. (If there were problems, a second unmanned flight could be made.) That was to be followed by a program of manned missions. Three orbital M-2s would be built. The mission profile was similar to that of the Dyna-Soar. The vehicle would be launched from Cape Kennedy into an elliptical orbit that would enter the atmosphere over Edwards. The pilot would fly the vehicle to a landing on the lake bed. The vehicle could be refurbished for another flight. Total cost of the program was between $150 million and $180 million.

We took the proposal to NASA headquarters, where it found enough support to obtain funding for two study contracts. NASA typically used this process to provide confidence in the feasibility and potential cost of a new concept. We then issued a request for proposals to industry and ultimately awarded one contract to McDonnell and one to Northrop. These were six-month studies intended to confirm our conceptual design or provide suitable alternative designs that could be developed and flown within our estimated budget. The studies went well. Both companies were enthused about the proposal and contributed company funds to enhance the quality of the final proposals, which were completed in January 1966.

Drake and I took the final study results produced by the companies back to NASA headquarters. Al Eggers was at headquarters at that time. With him as our advocate, we presented our study to a headquarters official. He listened attentively, asked a number of questions, and finally appeared to be convinced that we could successfully orbit and recover an M-2 lifting body. Following the presentation, we continued to discuss the probability of success. In our optimism, we might possibly have minimized the challenges associated with a mission of this type.

We were stunned when the headquarters official said, "If it is so simple and straightforward, why should we do it?" We tried to backtrack somewhat to admit that there were some real challenges involved, but we continued to insist that it was doable and that we would make a contribution to aeronautical and space technology by conducting the mission. We didn't convince him. We left NASA headquarters and returned to Edwards wondering what we had done wrong. Why did our brilliant idea backfire so abruptly?

By the summer of 1967, Northrop had come up with a better idea to demonstrate a lifting entry using the M2-F2. The new proposal would involve carrying the modified M2-F2 in the Apollo adapter module on a Saturn 1B launch with an Apollo command module. The mission could be a dedicated mission to launch the M-2 in orbit, or it could be an operational mission with the M-2 as a piggyback experiment.

There were several advantages to this proposal. First, we would not have to modify the Titan 2 booster to accept the M-2 as its nose cone, eliminating the need for an unmanned mission to verify that the combination was structurally sound and controllable. Second, the test pilot would be in the Apollo command module during the boost phase and would enter the M-2 only after reaching orbit. The test pilot on entering the M-2 would check out all his systems and only then separate for reentry if all systems were working normally. It was a much safer way to do a manned orbital mission.

A dedicated mission, devoted entirely to launching the M-2, would be more expensive than a piggyback mission. A Saturn booster was substantially more costly than a Titan 2, but the Saturn booster would not require any modifications to carry the M-2 into orbit in the adapter. It may have been a push moneywise.

Dale Reed remembers that Wernher von Braun was a big supporter of the idea. The German rocket engineer was at a meeting at Dryden when Reed proposed the idea of flying a lifting body aboard an Apollo mission to him. Braun told Paul Bikle, "I'll go ahead and prepare the rockets. You get the lifting bodies ready." The overall proposal was attractive to me, but I was not convinced that NASA headquarters would buy it. The proposal wasn't pushed too hard.

I did realize that the thermal protection materials for a lifting-entry spacecraft were not the optimum for refurbishment. The primary mission envisioned for an orbital lifting body would be transporting crewmen back and forth from a space station. The vehicle would have to be reusable. After landing, the heat shield would be refurbished for another launch. The McDonnell study proposed two different heat-shield designs for the orbital M-2. The first would have the nose, sides, bottom, and fin leading edges covered with ablative material. The upper surfaces and sides of the fins would use heat-resistant metals. The second had ablative materials on the forward third of the vehicle and the bottom flaps, while the rest would be heat-resistant metals. The first was cheaper to build but had higher refurbishment costs and was heavier.

We had only a little experience with ablative heat shields that could be refurbished, and little of that was good. In preparation for Mach 7 flights, the X-15A-2 had been covered with a spray-on ablative heat-shield material. The hope was that after a research flight it could be quickly and cheaply refurbished. In reality, the spray-on X-15A-2 heat shield was hard to refurbish, as the ground crew had difficulty ensuring that it was the proper thickness.

One question that was never fully answered was the effect of the rough ablative heat shield on the lift-to-drag ratio. With the suborbital X-23A unmanned vehicle, the added drag from the charred heat shield caused a 30-percent reduction. A lifting body needed all the lift it could get. The loss of lift from the heat-shield drag could have made landing marginal.

The breakthrough that made lifting reentry possible was developed by Lockheed with the help of Ames. This was a lightweight ceramic tile that could be bonded directly onto the skin of a vehicle. It was far more efficient and durable than the heat-resistant metal of the Dyna-Soar and did not have the drag and refurbishing problems of an ablative coating. These tiles cover the space shuttle. In the mid-sixties, however, they were "unobtainium."

SIMULATED NIGHT LIFTING-BODY LANDINGS

Lifting bodies had been proposed for use in space primarily because they could land on land. This eliminated the need for a large recovery fleet and potentially improved reuse and refurbishment, as the vehicle would not be exposed to saltwater contamination. Many other advantages of lifting bodies were proposed; however, some were pure conjecture, and others were not proven to be necessary.

We were being asked to demonstrate such functions as night landings, instrument approaches, ditching, runway landings, and arrested landings. Lifting-body supporters were trying to prove that these vehicles could do all those things and had the flexibility to be an operational spacecraft. A rendezvous with a space station might require a night launch and landing, and a system failure might force an immediate landing no matter what the time of day or weather conditions. It was, after all, night half the time. Opponents of the lifting-body concept, on the other hand, were asking that those capabilities be demonstrated before lifting bodies could be considered for use in space. The opponents in most cases were asking for more versatility than was available in capsule spacecraft or in any conventional aircraft.

In January 1967, I proposed testing the feasibility of low lift-to-drag landings at night. This did not sit well. In fact, I was accused of coming up with these dangerous programs after I quit flying. The purpose of the tests was not to satisfy either the supporters or the opponents. Rather I wanted to demonstrate an additional capability that would not normally be used operationally but that could be used if the additional risk was warranted to perform a mission. Adding night and instrument landings to our capabilities would improve mission reliability and ensure a normal recovery under any probable landing-site conditions.

Two different simulated lifting-body approaches would be made. One would use the main Edwards runway, with its normal runway lights plus an additional light to indicate the planned touchdown point. The other involved landings on the X-15 lake-bed runway. The C-47 would fly a racetrack pattern parallel to the runway and drop parachute flares. The F-104 would then land as the flares floated down. Two aircraft would fly the pattern; as one aircraft was touching down, the second was starting its approach. Both pilots would fly daytime simulated landings on the scheduled day of the night missions for practice and comparison with the night landings.

Bruce Peterson and Bill Dana were the first project pilots. The first night landing approaches were made on February 28, 1967, on the X-15 lake-bed runway, since it was assumed that any planned night landings would be made on the lake bed to take advantage of the large landing area. The results of these first night landings were so encouraging that plans for ground-controlled approaches were dropped as unnecessary in visual flight rules (VFR) conditions. Several of the touchdown points were on the desired location.

It quickly became clear that one flare was not adequate for the final touchdown because of a flicker effect. Two flares burning simultaneously were required for the final touchdown. To guard against the occasional dud, three flares were released from the C-47 about a minute before the first airplane was scheduled to land. Each flare illuminated a circle about 1.5 miles across and provided light for both landings.

Two more series of night landings were flown on March 7 and 21. The later date included the first supersonic approach. The initial conditions of altitude and speed were varied intentionally to duplicate the variations in arrival energy that could occur on actual missions. Two months after the final test, Peterson was injured in the M2-F2 crash. He had made a total of thirty-one simulated (day and night) approaches using F-104s. John Manke replaced Peterson for the remainder of the tests.

The night-landing program underwent a change in philosophy. It had been initiated at a time when the lifting-body flight tests were progressing satisfactorily, but after the crash, everything was reevaluated. The result was a decision to make the actual flights much narrower in scope than originally planned. At this time it was thought that a lifting-body vehicle might well have a landing engine, and therefore it would make an unpowered landing only in the event of a landing engine failure. Based on that, the comprehensive testing of unpowered night landings seemed unnecessary and would be terminated once the feasibility of night landings was established.

Only four more night tests were made, on August 10 and November 2, 7, and 8, 1967. The maximum error in all the tests was between 1,000 and 1,500 feet, more than adequate for a runway landing, and most were well within 500 feet of the desired touchdown location. A total of 47 night landings were made without any mishaps.

The pilots found that they could establish their location using the lights in and around the Edwards area. They preferred landing on the lake bed using flares to landing on the main runway with the normal runway lighting and the aircraft landing lights. The general view of the pilots was that unpowered lifting-body landings at night under VFR were feasible as a normal landing procedure for Edwards and might even be an acceptable emergency procedure for any 10,000-foot runway under VFR conditions, even if the pilot was not familiar with the local area.

One surprise for the pilots was the benefit of moonlight. The flights were made under conditions ranging from a full moon to no moonlight at all. The pilots found that even a quarter moon allowed them to distinguish the outline of the lake bed and see the black runway lines. The approach pattern was greatly improved, as they could see landmarks they used during the daytime flights. One of the pilots even made two lake bed landings in an F-104 without any other illumination than the light of a half moon. From the night tests we were all totally convinced that lifting bodies could land in almost any conditions of darkness and weather.

The night tests were also popular with my children. After a drop, I would go out and pick up the parachutes from the flares. I would then bring the chutes home for the kids to play with. On a windy day, they would hold onto the shroud lines and let the parachute drag them down the street.

Our faith in the possibility of a night landing was justified a decade and a half later. The first night shuttle landing was STS-8, which touched down just after midnight on Labor Day 1983. Lake bed Runway 22 was illuminated by six xenon arc lights, which put out 4.8 trillion candlepower. The landing lights shone across the lake bed, illuminating the runway markings much as the flares had on the earlier tests. Richard Truly and Daniel Brandenstein guided *Challenger* to a touchdown within 300 feet of the aim point. The second night landing at Edwards came two and a half years later, when *Columbia* touched down on January 18, 1986. The shuttles have no landing or anticollision lights, so they seemed to appear out of the darkness only seconds before touchdown.

THE SHUTTLE SHAPE DEBATE: BALLISTIC CAPSULE OR LIFTING ENTRY

There was a controversy within NASA in the late sixties over which should be developed first, a space station or a "shuttle vehicle logistics spacecraft" (or more simply, a space shuttle). Those supporting the space station first were proposing to build a capsule-type reusable spacecraft to carry people and supplies to and from the space station. Those favoring building the shuttle were advocating a fully reusable lifting-entry vehicle.

This split in post-Apollo goals was reflected in two competing design requests issued in 1968. The first, issued on June 17, was for a capsule-type spacecraft. A month later, McDonnell Douglas submitted a proposal to NASA to develop an enlarged version of the Gemini capsule for space station support.

The Gemini capsule was extended back to accommodate a crew of nine, and a cylindrical cargo module was added to carry 6,630 pounds of supplies. Called the Big Gemini, or Big G, it was intended to provide a space shuttle capability at low cost, using technology first developed for the Air Force's Manned Orbiting Laboratory (MOL) program. The Big G would use the MOL's Titan 3M booster, while the capsule would be based on MOL's Gemini B capsule. An enlarged follow-on version could accommodate up to twelve crew members and additional cargo. The Big G was strongly advocated by many people within NASA as the first space shuttle vehicle.

There was a serious drawback, however. After the mission was over, the Big G would use a paraglider and a three-skid landing gear to return to land. I had flown the Paresev, and a paraglider had been proposed as the recovery system for the Gemini capsule. Both had suffered a lot of problems. The testing of the Gemini paraglider had been somewhat successful, but there remained serious doubts about the practicality and reliability of such a system. The Big G capsule would have been larger and heavier than the Gemini capsule, and that would have only made the problems greater.

The second request was issued jointly by the Manned Spaceflight Center and the Marshall Spaceflight Center on October 30, 1968. Calling for an eight-month study of an "Integral Launch and Recovery System," it focused on various designs for lifting-entry spacecraft able to accommodate a crew of twelve and cargo between 5,000 and 50,000 pounds, in a cargo bay as large as 15 by 60 feet. The designs were broken down into Class 1, a reusable orbiter launched on an expendable booster (much like the orbital M-2, but on a larger scale); Class 2, in which the orbiter used expendable drop tanks; and Class 3, a fully reusable two-stage design.

In February 1969, four contractors were selected. McDonnell Douglas proposed a Class 1 vehicle using a lifting body with a crew of ten. Lockheed and McDonnell Douglas both proposed Class 2 vehicles: the Lockheed design was a long delta orbiter with a V-shaped tank called Star Clipper, while the McDonnell Douglas proposal used an orbiter with two large and two small tanks. The remaining three designs were all Class 3 vehicles. Convair's design used an orbiter with either two or three fly-back reusable boosters. Max Faget of the Manned Spaceflight Center submitted a design using straight wings for both the orbiter and booster, and a design from the Langley Research Center used an enlarged HL-10 as the orbiter. I was most involved with the final two proposals.

When Langley began developing the HL-10 shape in 1962, Gene Love envisioned that an operational vehicle would be between 25 and 30 feet long and carry a crew of twelve. In 1966 and 1967, Langley and Martin examined various proposals for crews of between one and eight astronauts, considering the effect on size, cost, and research utility.

For the October 1968 request, Langley simply scaled up the HL-10 research vehicle we had been flying at Dryden. The shape of both vehicles was identical. The HL-10 orbiter was 130 feet long and had a 15-by-60-foot cargo bay able to hold up to 50,000 pounds. Five 250,000-pound thrust rockets were placed beneath the center fin, while two jet landing engines were mounted in the base of each fin. (All of the designs carried jet engines for powered landings.) The Langley study noted that a fuel supply for one hour of subsonic cruise carried an expensive weight penalty. The HL-10 orbiter would be launched by a delta-winged manned booster about twice its length. There was now a battle pitting the Big G against the lifting-entry designs, and between the HL-10 orbiter and Max Faget's straight-winged orbiter.

The various designs were subjected to a period of analysis. It was during this time that I received a call from Gene Love asking if John Manke and I could meet him at NASA headquarters to attend a meeting with George Mueller, who was the associate administrator for manned spaceflight. Love wanted to convince Mueller that the HL-10 configuration was the most promising candidate for the shuttle orbiter. Of the many configurations under consideration, the HL-10 was the only one that had been successfully flown and landed by a pilot. The HL-10 had good handling characteristics and was the best of the three lifting bodies we had flown up to that time. It had a subsonic lift-to-drag ratio above 4 to 1 and no obvious aerodynamic problems, after the tip fins had been modified to eliminate the flow separation problems at high angles of attack.

John Manke had made the majority of flights in the HL-10 and therefore was the best qualified to describe its superior flying qualities to Mueller. Manke and I flew the red-eye flight to Washington the night before the meeting. We arrived at Dulles Field about 5 A.M. and boarded the bus for downtown Washington, D.C. We arrived at NASA headquarters just before 7 A.M. and had breakfast in the cafeteria. Our meeting with Mueller was scheduled for 8:30 A.M.

After breakfast we rode the elevator up to the sixth floor and then went into the men's room to clean up a bit. I shaved, washed my face and hands, combed my hair, and was ready to go to the meeting. I normally didn't bring a change of clothing with me on one-day trips to Washington. I had made many of these trips since I moved back to the research engineering organization: fly the red-eye into Washington during the night, attend the meeting during the day, then catch the evening flight back to Los Angeles and drive back to Lancaster, all in twenty-four hours. It was a grueling trip, since I couldn't get much sleep flying into Washington, but it was better than spending a night in the capital, in my opinion. I had begun to hate Washington, after so many trips to NASA headquarters. In fact, I began to hate traveling, period. Manke was relatively new to the traveling circuit. He had brought a change of clothes that included a new suit. He really looked sharp when we walked into Mueller's office.

Gene Love did most of the talking. There were only four of us in the office: Gene Love, John Manke, Dr. Mueller, and me. Love described the merits of the HL-10 and showed Mueller a sample of the wind tunnel data. When Love completed his presentation, he asked Manke to discuss the flight characteristics of the vehicle. Manke gave an enthusiastic description of the vehicle's flying qualities and narrated a short film showing the HL-10 in flight.

Manke gave an excellent sales pitch touting the virtues of the HL-10, and Mueller was obviously impressed. He was not convinced, however. Mueller had previously heard pitches from other advocates of competing configurations. Those advocates had highlighted their competitors' deficiencies, so Mueller was primed to challenge Love on the merits of the HL-10. One of the disadvantages of the HL-10 was that it couldn't easily accommodate a 15-by-60-foot payload bay. If the HL-10 was scaled up to accommodate a payload bay of that size, the overall vehicle size would be excessive, compared with some of the other vehicles.

Another disadvantage was its highly contoured aerodynamic shape. Any propellant tanks in the vehicle would have to be integral tanks rather than the structurally more effective cylindrical tanks. Integral liquid hydrogen tanks would be difficult to seal and also prone to leaking after undergoing numerous fueling and operational cycles. Mueller wasn't convinced that the HL-10 was the best configuration for the shuttle. He did agree with Gene Love that the shuttle should be a lifting-entry vehicle that could maneuver in the atmosphere and land horizontally like an airplane. We at least scored a victory in the battle between the lifting-entry configurations and the ballistic capsule configurations. The Big G faded from consideration.

We may not have been solely responsible for Mueller's decision to go with a lifting-entry configuration, but we had made a convincing argument and claimed some credit for the victory. Future astronauts would return from space in a much more dignified manner than their predecessors who had unceremoniously splashed down in the ocean. They would step out of their space plane and walk down the ladder to dry land, a fitting ending to an exciting space voyage.

When Manke and I came to work the next day, the other pilots asked how the presentation was received by Mueller. Tongue in cheek, I told them that I had overheard Mueller telling Love that he didn't understand much of Manke's presentation, but he thought that Manke was a pretty snappy dresser. Manke didn't live that one down. We still taunt him with the "snappy dresser" label.

THE HYPER 3: UNMANNED VERSUS MANNED RESEARCH VEHICLES

In December 1969, I again found myself at the controls of a lifting body. This time, however, I was flying it from the ground. When work started on the heavyweight lifting bodies in 1964, Dale Reed was named head of advanced planning at Dryden, which gave him the freedom to develop new ideas without having to face the day-to-day problems of program management. During the early sixties, he had tested a series of lifting-body shapes by building small models, then launching them from a large radio-controlled model airplane. Called *Mother,* it had a 10-foot wingspan and was powered by two engines. The lifting-body models were attached to the bottom of *Mother,* then released at about 1,000 feet above the lake bed. By late 1968, *Mother* had made more than 120 drops.

Following Mike Adams's death in the X-15, we installed an eight-ball attitude indicator in the control room. This was to show the ground controllers the vehicle's attitude in flight. Reed and I were monitoring a flight when he asked me if a pilot could actually fly a research vehicle using the eight-ball. I said I could. A month later, at a cost of $500, I was flying *Mother* from the ground using the eight-ball. It was like flying on instruments. Next, Reed wanted me to try it with a real research vehicle.

Reed had flown some models of a Langley-designed lifting-body shape called the Hyper 3. It was similar to the FDL-7 and the Lockheed Star Clipper: a long delta-shaped fuselage, two vertical fins, and stub wings. Long, straight wings would extend out from the fuselage to increase the lift-to-drag ratio for landing. The original lifting bodies could land up to 1,000 miles off their ground track, but the Hyper 3 had a cross-range capability of 4,000 to 5,000 miles. Reed selected this shape to build into a full-scale unmanned lifting body.

The Hyper 3 consisted of a framework of steel tubing covered with Dacron fabric and aluminum and fiberglass panels. It was 31 feet 8 inches long, weighed about 1,000 pounds, and cost $6,500 to build. We did have hydraulic controls on the vehicle. I controlled it in flight using a ground cockpit with a full instrument panel.

On December 12, 1969, the Hyper 3 was ready for its first flight. The vehicle was attached to a 400-foot cable and carried to 10,000 feet by a helicopter. Problems immediately became apparent; the vehicle was unstable on the lift. In fact, it was damn near in a stall when we launched it. I quickly found out that it was hard to land a model. Working from the outside in just takes a lot of adaptation. I flew it for about 3 miles, then I reversed its course and flew it another 3 miles. Initially, there had been some thought that I might be able to land it myself, but I was effectively flying on instruments, and you just can't land on instruments. It ended up that Richard Fischer, who was used to landing models, took over about 1,000 feet in the air. He rolled the Hyper 3 slightly to get a better look at the vehicle, and then landed it. Flying the Hyper 3 was just as demanding as flying a manned research vehicle. I have never come out of a simulator as emotionally and physically drained as I was after the Hyper 3 landed, even though the flight was only 3 minutes long.

That was the Hyper 3's only flight. The vehicle had a much lower lift-to-drag ratio than had been predicted. Dryden was also working on the manned lifting bodies, the YF-12A program, and the Supercritical Wing F-8 at the time. They had higher priority than a little shoestring operation being done in-house. Nevertheless, the Hyper 3 was only the first in a series of unmanned research vehicles flown at Dryden during the seventies and eighties. A ⅜ scale model of the F-15A Eagle was stall and spin tested, the Himat tested a number of new aerodynamic technologies, the Drones for Aerodynamic and Structural Testing (DAST) tested flutter suppression technology, and the Mini-Sniffer was a prototype for an aircraft able to fly in the atmosphere of Mars.

These examples raise the issue of manned versus unmanned research vehicles. A manned vehicle gets more program visibility because the pilot's handling comments tend to draw news-media attention and technical interest. With an unmanned vehicle more risks can be taken, providing management is willing to accept the loss of the vehicle and the experiment. If the experiment becomes too expensive, the advantages of an unmanned vehicle quickly diminish. The loss of the vehicle becomes unacceptable, and the vehicle becomes more complex to ensure survivability. In such cases, a manned vehicle can be simpler than an unmanned vehicle, since the pilot can be used to provide redundant recovery options.

An unmanned vehicle can be made somewhat cheaper by eliminating life support systems and the need to man-rate the vehicle. An unmanned vehicle may be significantly cheaper if elimination of the pilot substantially reduces its size and complexity.

An unmanned vehicle is justified regardless of cost or other considerations if the experiment is so hazardous that it cannot be safely accomplished in any manner with a manned vehicle. The DAST program successfully tested a state-of-the-art flutter suppression system, even though the vehicle was lost in the process. The loss of the vehicle during a test of the system lends credence to the choice of an unmanned vehicle to conduct this type of hazardous testing. The reality is that few experiments meet those criteria. When old test pilots get together to swap stories about their close calls, they don't talk about research flights but rather about cross-country and proficiency flights. Oddly enough, many of the fatal accidents at Edwards had nothing to do with flight testing.

I saw a tragic example on March 1, 1976. I was about 10 miles southwest of the main base when I noticed a black pillar of smoke up ahead. In the early morning, smoke usually rises in a narrow column in the desert; the winds are generally calm, and there is no turbulence to disperse the smoke. A smoke column is visible 40 to 50 miles away in the clear desert air, and since the desert is relatively flat and treeless, one might even see the fire itself at that distance. Smoke in the desert is usually man-made, since there is little vegetation to burn naturally. I dreaded seeing a column of smoke like that up ahead. After twenty years at Edwards, I knew that it usually marked an aircraft accident. As I got closer, I saw that the smoke was rising from a fire on the north lake bed. It quickly became obvious that it was an aircraft accident. Crash and rescue and fire trucks with their red lights flashing surrounded the crash site. From the road above Dryden, I had an excellent view of the accident scene. The fire was still smoldering, but the wreckage was covered with foam.

As I walked into my office, someone told me that Mike Love was the pilot of the plane. Witnesses on the ground saw the RF-4C aircraft trailing smoke as it approached the lake bed and then burst into flames. One occupant ejected and descended by parachute. The aircraft continued to descend toward the lake bed, burning furiously. It finally hit in a huge ball of fire. Crash and rescue vehicles swarmed onto the lake bed in response to a Mayday call before the crash. They quickly attacked the fireball on the lake bed and within minutes had eliminated the open flames. One rescue vehicle picked up the aircraft occupant who had ejected. It wasn't Love. It was the backseater, the test engineer. Love was still in the cockpit when the aircraft hit the lake bed.

Subsequent questioning of the test engineer revealed that the aircraft cockpit had filled with smoke from the fire. The smoke was so bad that he had pressed his face against his panel to read the instruments. Love pulled the ejection seat handle, and the test engineer felt himself being jerked back by the straps. The test engineer's canopy separated and his seat ejected. Love's canopy and seat failed to separate because of fire damage, and there was not enough time for him to complete the emergency procedures before the airplane rolled over and hit the lake bed.

Love had been the Air Force program pilot for the X-24B lifting-body program. He had completed two glide flights and ten powered flights, including the second runway landing of a lifting body. Love had done all that, yet he died on an ordinary proficiency flight.

ELIMINATION OF THE SHUTTLE LANDING ENGINES

The lifting-body program provided us some legitimacy and credibility in our discussions on the shuttle. Dryden management was convinced that the shuttle could and should be landed without power to eliminate the proposed landing engines. The engines and the fuel required to make a powered approach would significantly increase the weight of the orbiter and the liftoff weight of the total shuttle vehicle. I showed some films of our lifting-body landings during a trip to the Manned Spaceflight Center in Houston. The films were impressive. They were taken from about half a mile away from the runway to enable the photographer to capture the steep approach, the flare maneuver, and the landing rollout. At those distances, the lifting bodies appeared small, almost like small birds.

The lifting bodies appeared to be birds of prey swooping down to make a kill. Max Faget, the director of engineering and the chief spacecraft designer at Houston, commented that the lifting bodies were "a stunt perpetrated by those hot test pilots at Edwards." Faget did not like the low lift-to-drag ratio of either a lifting-body shape or a delta wing design. His straight-wing shuttle design had a good subsonic lift-to-drag ratio. The lifting-body landings were quite spectacular, starting from a 20-to 30-degree dive at 300 knots and ending with a 200-knot touchdown in less than 25 seconds elapsed time. I had to admit that the landings looked much worse from the outside than from inside the cockpit. I knew we had a big sales job to do to persuade anyone to take unpowered landings seriously for normal operation of the shuttle.

One reason the landings appeared to be a spectacular stunt was the optical illusion created by such a small vehicle flying through the maneuver at high speed. A larger vehicle would make the same maneuver look much more leisurely. During the trip back to Edwards, I was trying to determine how best to demonstrate that, but I couldn't think of a good way to do it.

During a postflight debriefing of an HL-10 flight, I overheard Fitz Fulton describing the way he followed the HL-10 down in the B-52 during the landing approach. He indicated that on many HL-10 flights, he would head straight in for the landing runway after launching the HL-10 and be in position to follow the HL-10 down to a landing. I asked him if he could descend as fast as the HL-10. He said he could by dirtying up the B-52, that is, extending the landing gear, deploying the wing spoilers, and reducing thrust to idle. I then asked if he would be prepared to demonstrate that capability on our next scheduled launch. Fulton said he would.

In preparation, we made plans to track the B-52 and record airspeed versus altitude during the descent. I also requested that motion pictures be taken of the B-52 during the final approach and landing from several different locations. After our next HL-10 flight, Fulton made several simulated unpowered approaches. Following the flight, we examined our data and found that the B-52 could achieve some surprisingly low lift-to-drag ratios during the simulated shuttle approaches. Fulton was right; he could follow a lifting body down reasonably well. Our films also proved our point about the illusory effects of a large airplane. The simulated unpowered landing appeared to be a benign maneuver when performed by a B-52.

We now had to demonstrate that the landing maneuver was not a stunt. To do this, I invited the shuttle program manager, Bob Thompson, to fly as a passenger in one of our F-104 aircraft that we used to simulate and practice lifting-body landings. He accepted the invitation. He was exposed to a series of simulated unpowered lifting-body approach and landing maneuvers from inside the F-104. He was impressed by the gentle nature of the maneuver and, as a result, became a believer.

On a subsequent trip to Houston, I showed the movies of the B-52 making simulated unpowered shuttle approaches and described the extensive experience gained making routine landings in various rocket aircraft at Edwards. At that time, we had accumulated a total of almost 600 unpowered landings at Edwards in the X-1s, X-2, D-558-IIs, X-15s, and lifting bodies. In all that experience, we had no accidents attributable to the unpowered nature of the approach—quite an impressive accomplishment. The resistance to unpowered landings began to soften among the shuttle program participants.

We followed up our advocacy efforts with more F-104 rides for program managers and a follow-on large aircraft program to demonstrate further the feasibility of unpowered approaches in the shuttle. I initially wanted to borrow or lease a DC-8 for more demonstrations, since the DC-8 thrust reversers could be deployed in flight. In-flight thrust reversal could substantially reduce the aircraft's lift-to-drag ratio to that of the shuttle orbiter. We were unsuccessful in quickly acquiring a DC-8, however. Fred Drinkwater, a NASA Ames pilot who had developed a high-speed unpowered landing technique, suggested that the Ames CV-990 might be capable of matching the orbiter's lift-to-drag ratio during simulated unpowered approaches. He brought the airplane down to Edwards, and we flew a short test program to measure its performance. It appeared capable of simulating a shuttle orbiter during landing approaches.

We formally requested Ames Research Center support to demonstrate simulated unpowered approaches to various shuttle program participants. We requested that several astronauts be designated to fly demonstration flights, with Fred Drinkwater flying as instructor pilot. We also encouraged Houston to send other observers to ride as passengers. These demonstrations were successful in minimizing Houston's resistance to unpowered landings.

We invited two airline pilots to fly the 990 during simulated shuttle approaches, and they did an excellent job on their first flight. We challenged them further by simulating instrument flight rules (IFR) conditions during a major portion of the approach. They again did a great job. When they finished their demonstration flights, they were staunch supporters of unpowered landings. They were so enthusiastic, we were afraid they might try some simulated unpowered approaches during scheduled airline flights; they had threatened to do just that.

Another issue regarding unpowered landings was whether an astronaut would be proficient enough after an extended stay in orbit to make an unpowered landing. We normally practiced extensively before a real unpowered landing, including practice landings the morning of an X-15 or lifting-body flight. To address this issue, we proposed that two noncurrent pilots fly approaches in the 990 without any practice. Bruce Peterson and I were selected as the test subjects for this demonstration.

Peterson had not flown either a real or simulated approach in over six months. I had not flown any airplane for almost two years. We both flew with Fred Drinkwater on the same flight. Peterson touched down within 750 feet of the desired point on his first approach. I touched down within 1,000 feet of the desired point on my first approach. Based on that limited sample, we concluded that an astronaut could land quite accurately even though he had remained in orbit for an extensive period of time. Houston was beginning to believe that unpowered flight might be an acceptable mode of operation.

The Air Force ran its own program of simulated lifting-body landings in an F-111A. In June and July 1969, six F-111 flights were made by Peter Hoag. The last flight ended with the loss of the landing gear doors. With the flying phase only half-finished, it was put into a state of limbo due to paperwork problems. This continued until September, when Hoag flew Gen. Alton Slay, the Edwards base commander, on several approaches in the F-111. General Slay's impression was that it wasn't as hairy as he had expected. Approval to resume the tests was given the following week, and several flights had been made by mid-September 1969. Tests continued until November 10, which included IFR landings and night landings.

By early 1970, the tests were complete, and Dryden sponsored a symposium to present data from our various flight programs that we felt were pertinent to the shuttle. We presented papers on the results of the lifting-body program, the B-52 and 990 landing program, and the Air Force F-111 tests. We invited managers from Marshall and Kennedy space centers as well as NASA headquarters. The symposium was well received.

THE HL-10 POWERED LANDING TESTS

There was one final test of the feasibility of a powered shuttle landing. In February 1970, the HL-10 was grounded. The XLR-11 engine was replaced by three hydrogen peroxide emergency landing rockets, each with a thrust of about 300 pounds. A large propellant tank was also installed. The low-thrust engines were to be used during landing to produce a less steep, more airplane-like angle. With the rockets firing, the approach angle was cut from 18 degrees to only 6 degrees.

The plan called for Peter Hoag to maintain an airspeed of 280 knots with the rockets burning. At 200 feet above the lake bed, the pilot would shut down the rockets, decelerate, and flare to a landing. Two powered approaches were made, on June 11 and July 17, 1970. The results sealed the fate of the shuttle landing engines. Hoag found that the shallow approach angle made it hard to judge the touchdown point, while the nose-high attitude required that he look through the nose window, with its distorted image. A powered landing actually increased the pilot's workload over that of a normal unpowered landing.

The biggest problem, however, was the question of what to do if the landing engines did not start. The powered landings both used Runway 17, which stretched 7 miles across the lake bed. As the HL-10 glided down, the pilot set up an unpowered approach on the near end of the runway. If the rockets failed to fire, he would simply continue the approach to a landing. Once he had the rockets firing, he established a new aim point about 4 miles down the runway. Such a procedure was not feasible with a normal-sized runway.

At the 1970 symposium I summed up the landing-engine situation this way: "[Our pilots] would refuse to rely upon them to make a successful approach and landing. The shuttle, whether it has landing engines or not, must be maneuvered, unpowered, to the point near the destination because the engines cannot be started until the vehicle is subsonic and only limited fuel will be available. To us it seems ridiculous to maneuver to a position where power must be relied upon to reach the runway. Instead, we would maneuver to a high key position to begin an unpowered approach. Then, regardless of whether the engines could be deployed, started, and kept operating, a successful approach and landing could be made."

Given such conditions, however, the landing engines themselves were redundant. Since the vehicle had to fly an unpowered approach, it made no sense to then switch to a powered profile, with its added problems. Unpowered shuttle landings became a program baseline position. Landing engines were eliminated.

THE SPACE SHUTTLE APPROACH AND LANDING TESTS

The elimination of the landing engines did raise another problem, the question of moving the orbiter around the country after landing. It would not be a problem if the orbiter landed at the launch site at the conclusion of each flight, but everyone realized that the orbiter might have to land at some other site because of weather, an emergency, or a change in mission plans.

The HL-10 orbiter design had a large straight wing fitted into the payload bay, with its own engines and fuel supply. Once the space shuttle design was finalized, several ideas were bandied about: an airship, a huge helicopter, an air-cushion vehicle, or a C-5 transport that would tow the orbiter behind it (shades of the C-47/M2-F1 flights). The only idea seriously considered initially was a set of strap-on jet engines for use during ferry operations. That was certainly a viable option, but the range of the orbiter with strap-on engines was severely limited (less than 500 miles) by its low lift-to-drag ratio, which necessitated high-thrust engines and high fuel usage. The resulting limited range would require four or more stops to fly from Edwards to the Kennedy Space Center, an undesirable operation with such a complex and expensive vehicle.

Paul Bikle suggested a mothership when I discussed the problem with him. I had also thought of a mothership but had rejected it, daunted by the massive size and weight of the orbiter. I didn't believe that any existing aircraft could carry it into flight. It was projected to weigh more than 200,000 pounds and be over 120 feet long. The other major problem was again the orbiter's drag. The extra drag of the orbiter would require a lot of excess thrust to get airborne and stay airborne.

Bikle thought that an existing aircraft such as the 747 or the C-5 might be able to do the job. I was not convinced. I did suggest the idea at the next meeting of the Space Transportation System Technology Steering Committee. Some of the committee members felt the idea had merit. I was still somewhat skeptical, but we began working with Houston to develop a test program to demonstrate the practicality of launching the orbiter from the 747. Dryden was given the responsibility of demonstrating that the modified 747 could take off and fly with the orbiter mounted on top. We did a limited test program, then began the approach and landing test (ALT) program.

Between August and October 1977, five successful glide flights of space shuttle *Enterprise* were made following launches from the 747 shuttle carrier aircraft. The ALT program proved me wrong. Top launches were quite safe. The orbiter rose up off the 747 so smoothly and majestically that it appeared to be in slow motion. It was an amazing sight. The ALT program was a real success. Three flights were made using a tail cone on the orbiter that reduced the base drag and increased the lift-to-drag ratio enough to minimize the rate of descent and simplify the landing task. Two more flights were made without the tail cone to verify that successful landings could be made after a return from orbit. All of the landings were successful, although not without incident. On the final flight, which was an attempt to land on the Edwards main concrete runway, there was a pilot-induced oscillation just before touchdown.

The shuttle program received a real boost in confidence and visibility as a result of these tests. It had been fourteen years since I had first taken the little wooden M2-F1 aloft behind Whitey Whiteside's Pontiac. Now astronauts were ready to go into orbit aboard a lifting-entry spacecraft.

11
LIFTING BODIES RESURGENT

This chapter was written entirely by Curtis Peebles.

It was not until the final years of Milt Thompson's life that the lifting-body concept staged a resurgence. The lifting bodies had influenced the design of the space shuttle in several ways. The HL-10 powered landing tests had shown that landing engines were impractical. A more basic contribution was speed brakes. The HL-10 used a split rudder as a speed brake, and the X-24A and M2-F3 used their tip fin rudders in a similar way. The pilot used them like a throttle, opening them to increase drag and closing them to maintain the vehicle's energy state. The lifting-body experience showed that speed brakes were absolutely mandatory. Based on this, the space shuttle was fitted with a similar split rudder.

The final configuration and shape of the space shuttle were influenced by both technical-cost and mission-political factors. The original 1969–1970 idea of a two-stage, fully reusable design proved both too difficult and too costly in the post-Apollo era. The fly-back manned booster was replaced with a disposable external tank and two solid rocket boosters. The external tank would be released and burn up, and the boosters would be fished out of the ocean and reused.

The orbiter's design was influenced by Air Force mission needs and NASA political requirements. The political realities of the early seventies required the shuttle to replace all existing expendable boosters. That meant NASA needed the Air Force as a customer. What the Air Force needed was a 15-by-60-foot payload bay and a high cross-range capability. To supply the first, the shuttle was built around a large payload bay. Although a straight-wing design could probably have been built to meet the Air Force cross-range requirement, the shuttle has delta wings. The problems with straight wings were that they were difficult to protect from reentry heating and they caused interference heating of the horizontal tail. The delta wings gave the shuttle a configuration more like that of the Dyna-Soar.

THE SOVIET CONNECTION: FROM SPIRAL TO URAGAN

The road to the resurgent lifting-body program led through Moscow and was influenced by the loss of space shuttle *Challenger.* Like the United States, the Soviet Union had studied lifting-entry spacecraft as early as 1958, and like the Dyna-Soar, the effort suffered from political difficulties and internal rivalries. Twice, in 1960 and 1964, development programs were canceled. In 1966, the Spiral project was begun. The Mikoyan-Gurevich design bureau vehicle had a hypersonic first stage that carried on its back an expendable rocket and the Spiral spacecraft. The shape was like a lifting body, similar to the HL-10, but with a raised cockpit. During reentry, the tip fins were angled up, like the HL-10. During landing, they would be lowered, becoming wings. The Spiral's mission was satellite interception and destruction. Mission duration was to be a mere three orbits.

In 1968 and 1969, three subscale Spiral models were launched on suborbital flights from the Plesetsk Cosmodrome in the northwestern USSR. They landed at the Kapustin Yar missile test center, some 2,000 kilometers away. They were called BOR-1 to -3 (an abbreviation of Becpilotniye Orbitalniye Raketoplan, or unpiloted orbital rocketplane). The plan was for unmanned full-scale orbital tests in 1970, with the first manned flight of the two-stage vehicle to be made in 1977.

The next step was to conduct manned subsonic and supersonic flights within the atmosphere, much like the NASA lifting-body flights of 1966–1975. These were canceled in 1969, before they could begin. Reasons vary as to why. One version blamed the cost of the Soviet manned lunar landing program, while another pointed to technical difficulties with the hypersonic first stage. Still another quotes Soviet defense minister Gretchko as saying, "We will not engage in nonsense."

The Spiral atmospheric test program was revived in 1972, and between May 1976 and September 1978, eight manned flights were made. The vehicle, called Article 105.11, made both ground takeoffs and air drops from a Tu 95 bomber. Jet and rocket engines were tested, as was a skid landing gear. On one hot summer day, the skids sank into the black-topped surface of a runway. The engineers smashed watermelons over a 70-meter stretch of runway, and put watermelon halves under the skids. So lubricated, Article 105.11 successfully took off. Word of the vehicle spread in the West and was the first indication that the Soviets were undertaking a shuttle program.

While Article 105.11 was being flown, there were not one but two shuttle programs under way in the USSR: Buran (Snowstorm) and Uragan (Hurricane). Buran was a clone of the U.S. space shuttle. The orbiter was nearly identical in shape, and it used a booster configuration with an external tank and strap-on boosters. In Buran's case, there were four strap-ons rather than two, and the fuel was liquid rather than solid as with the shuttle. The main engines were also on the external tank, rather than on the orbiter.

There was one substantial difference between the space shuttle and Buran. The U.S. space shuttle was designed as a low-cost means of delivering satellites into orbit. The Soviets deluded themselves into thinking it was really a space bomber, able to swoop down on Moscow, drop a nuclear weapon, then fly away. Buran was to be the Soviet counterpart to the imagined shuttle bomber, but the Uragan program was to build a shuttle killer. This space fighter was a two-man spacecraft based on the original Spiral shape. It was to be armed with a recoilless gun and was intended to destroy shuttles launched into polar orbits from Vandenberg Air Force Base. The Soviets believed that such missions would make the entire Soviet landmass vulnerable to nuclear bombing from space. The Uragan was to be launched on the Zenit booster and be operational in the late eighties.

As with the similar Spiral, subscale models of the Uragan underwent reentry testing. The first launch of the BOR-4 vehicle was made on June 3, 1982. Cosmos 1374 made an orbit of Earth, then descended by parachute into the Indian Ocean. Waiting for it was a Soviet recovery fleet and a Royal Australian Air Force P-3 patrol plane. As the vehicle was brought aboard the recovery ship *Yamal,* the P-3's crew photographed the proceedings. Another three subscale tests had been flown by December 1984. Yet, despite the apparent success, no manned Uragan flights were ever made.

Both Uragan and Buran, like all the other Soviet lifting-entry spacecraft programs, were canceled. Uragan ended in 1987, following the decision by the United States, in the wake of the *Challenger* accident, not to launch shuttle missions from Vandenberg. Buran made an unmanned two-orbit flight in November 1988. Although there was talk of manned flights in the near future, the project was eight years behind schedule and far from ready. In November 1991, the Soviet military pulled out of the project, and the formal cancellation was made in June 1993. Total cost of the project was estimated at $30 billion, about three times that of the U.S. space shuttle.

The 35-year secret history of Soviet attempts to build a lifting-entry spacecraft proved to be both a technical and personal failure. Between 1965 and 1989, nearly sixty cosmonauts were selected for the various projects, but none was ever to fly in them. Yet in February 1994, eight months after the end of the Buran program, a cosmonaut finally flew in a winged spacecraft. His name was Sergei Krikalev, and he was aboard space shuttle *Discovery*.

THE HL-20

Around 1987, there was a rebirth of U.S. interest in lifting bodies, reportedly an outgrowth of the BOR-4 flights. Published accounts state that in 1983–1984, Langley conducted classified studies of the BOR-4 for the Air Force. Using the photos taken during the recovery, these reports continue, a computer model of the vehicle was made by the Langley scientists, followed by wind tunnel and fluid dynamics tests.

At the same time, the loss of *Challenger* caused a change in crew safety plans for space station *Freedom*. Before the accident, a safe haven was planned, where the crew could take refuge in the event of a fire or system failure. If necessary, a shuttle would then be launched to bring them home. In June 1987 this was changed to a lifeboat that would be attached to the station, called the crew return vehicle (CRV). Three different concepts were looked at. First was a ballistic capsule, a larger version of those used by the unmanned Discoverer program, which had the problem of high reentry g loads of 7 to 8. The second was a capsule that developed a small amount of lift, like Apollo. This reduced the reentry to 2 to 3 g. There was even some talk of refurbishing Apollo capsules from museums for use as the CRV. The third possibility was a lifting-entry vehicle, with a reentry g-loading of 1 to 2. This was an advantage for certain medical conditions but required added complexity and cost over a capsule-type CRV.

In 1987, three years after the studies of the BOR-4, Langley proposed the HL-20, a lifting body able to carry two pilots and eight passengers. Although the HL-20 was intended to serve as a CRV, the vehicle could also serve as a space taxi, to take new crew members and supplies up to the station and then return with the old crew. The HL-20 launch vehicle was to be a Titan 4, the duration in orbit would be 72 hours or less, and the reentry g-loading would not be more than 1.5. Initial estimates were that it would take only 20 to 25 percent of the man-hours needed to prepare a shuttle for launch.

Langley explained that the HL-20 design was based on previous NASA and Air Force lifting-body programs. Others, however, noticed that it looked very much like the BOR-4 vehicle. Both have a similar-shaped fuselage, a small center fin, and two large tip fins. There are certain differences: the BOR-4 has a turned up nose and a flatter upper surface, while the HL-20 has a more rounded nose and sloping top.

As part of its research, Langley built an HL-20 simulator for landing trials. Bill Dana flew the simulator and found the vehicle handled as well as any of the earlier lifting bodies. The HL-20 had a lift-to-drag ratio of 4.2 to 1, on a par with the HL-10 and X-24A, which made it easy to land. The HL-20 also had the advantages of avionics designed for the shuttle, which had not existed in the sixties and early seventies. These included a heads-up display and an inertial navigation system.

Milt Thompson felt that a small lifting-body design had a future. He noted, "I think there is going to have to be a simpler and cheaper vehicle to move people back and forth than the shuttle." Thompson also noted that the situation in 1992 was different from the sixties. The time pressures of the space race favored simpler capsule designs. In the early nineties, the pressures came from budget limitations, which made a reusable design attractive. He concluded by saying, "Even if we didn't have a space station, I think the next space vehicle will probably be a lifting body."

By this time, however, history had repeated itself. The lifting body HL-20 was rejected in favor of a capsule as an interim CRV. But the world was also a different place from what it had been in the sixties. The capsule selected as interim CRV was a Russian Soyuz. It would provide a man-rated, fully automated vehicle for initial station operations, but with limitations. The Soyuz can accommodate only three people, in cramped conditions. One modification required was seating able to fit taller astronauts. Passengers must sit with their knees practically in their chest, an uncomfortable position for someone with appendicitis. The most basic problem is that when the station is operational, the crew will grow to six. The solution, as Milt Thompson foresaw, was a lifting body.

THE X-38 CRV: THE RETURN OF THE FINNED POTATO

The story of the X-38 CRV actually began three decades before, with the X-23A and X-24A programs. The high-speed testing had been done by the X-23A subscale vehicles in 1966 and 1967. Three suborbital flights were launched from Vandenberg Air Force Base by an Atlas booster. The Atlas flew a high, arching trajectory, and the X-23A would then separate and reenter the atmosphere. During the reentry, the vehicle would make its test maneuvers. After the heating phase of reentry was completed and the vehicle had slowed, a 48-foot-diameter parachute would deploy. Waiting below were three JC-130B recovery planes. Using lines rigged between two poles, a plane would catch the parachute and bring the vehicle aboard. Because the large parachute would not fit between the poles, a thimble-shaped air pickup cone was attached to the top of the chute.

The first X-23A launch was made on December 21, 1966. The launch simulated a reentry from low-Earth orbit, with no cross-range maneuvers. The performance of both the booster and the vehicle was excellent; the X-23A flew 4,300 miles down range and arrived within 900 feet of the aim point. Then failure struck—the main parachute failed to deploy, and the vehicle hit the Pacific Ocean at high speed. Despite the last-second failure, the mission was judged a success. It had shown that a lifting body could successfully reenter the atmosphere, that it was stable at hypersonic and supersonic speeds, and that actual flight behavior was as predicted.

The second X-23A launch was made on March 5, 1967. The vehicle successfully made a cross-range maneuver of 500 nautical miles by holding a constant bank angle of 46 degrees throughout reentry. This was the first time a vehicle had maneuvered during reentry. The vehicle arrived within 2 miles of the aim point, and parachute deployment was successful. One of the JC-130B recovery planes made a flyby of the descending parachute, but the X-23A was hanging nose down rather than nose up as expected, and the recovery was not attempted. The vehicle splashed down safely, but the flotation bag tore free and it sank. Despite the loss of the vehicle, virtually all the data from the flight were successfully received.

The third X-23A followed on April 19, 1967. The flight simulated a reentry from low Earth orbit with a maximum cross-range maneuver. During reentry the vehicle was maneuvered more than 710 miles cross-range to a point less than 5 miles from the intended aim point. Down below, three JC-130Bs of the 6594th Test Group waited. This unit made midair recoveries of reconnaissance satellite capsules. The first plane missed on its recovery try, and the second plane passed on the attempt. It was now up to the third plane and its pilot, Capt. Richard M. Scofield. He had already made one midair recovery, the first capsule from the Corona 112 reconnaissance satellite. He lined the JC-130B up on the air-pickup cone and snagged it. The X-23A was then reeled aboard the plane. Scofield would later go on to make a total of eight recoveries and become one of the unit's aces. The X-23A was found to be intact following its recovery. The program was so successful that a fourth launch was canceled.

Later that year, the manned X-24A was delivered to the Air Force and, in early 1969, began its first glide flights. They were followed by powered flights, reaching a maximum speed of Mach 1.6. Of the various lifting bodies, only the SV-5 shape had been tested from Mach 25 down to landing speeds.

When development work began on the CRV, it was important to use as much off-the-shelf technology as possible to reduce costs. That included not only subsystems but also the shape. The HL-20, despite its resemblance to the BOR-4 vehicles, did not qualify. That left the potato with fins. The X-38 is an enlarged copy of the X-23A, right down to the simulated bulge for the pilot's canopy. It is meant to test the design, systems, and recovery procedures for the operational CRV. At 28.5 feet long and 14.5 feet wide, it is only slightly larger than the X-24A (which was 24.5 feet by 13.66 feet). Like the X-23A, the X-38 lacks the center fin of the X-24A.

As a lifeboat, the CRV has an unusual operating profile. In the event the space station had to be evacuated, the six-member crew would climb aboard and separate from the station. The vehicle would orient itself in space using nitrogen gas thrusters, retrofire, then jettison its deorbit engine module. The same as the space shuttle, the CRV would glide toward a landing. The CRV's final descent would use a steerable parafoil parachute, deployed at 20,000 feet. Just before touchdown, skids would extend from the underside to act as landing gear. The vehicle would use satellite navigation data throughout the landing, so it could operate without a trained pilot.

Although the CRV will be automated, the passengers will have some control. They can control the vehicle's attitude in space, pick the landing site, switch to backup systems, and steer the parafoil if necessary.

The X-38 program began in early 1995 as a small in-house effort at the Johnson Space Center. Dale Reed, who had retired from NASA, was also involved, acting as a consultant. Reed quipped, "My official title is chief wizard." As with the original M-2 more than three decades before, his efforts began with models. In the summer of 1995, he took a four-foot-long, 150-pound X-38 model aloft in a Cessna. At an altitude of 10,000 feet, he pushed it out of the plane. It glided 2 miles, then deployed a parachute and landed.

Early in 1996, a contract was issued to Burt Rutan's Scaled Composites, of Mojave, California, for construction of three full-scale atmospheric test vehicles. The first was delivered to the Johnson Space Center in September 1996 for installation of avionics, computer systems, and other hardware. The second followed in January 1997. Existing technology accounted for as much as 80 percent of the components.

The X-38's flight computer and its software operating system are both commercial products already in wide use. The X-38's global positioning system and inertial navigation system are now in use as a unit aboard Navy fighters. The heat-shield coating had been previously developed by NASA and is much more durable than the space shuttle tiles. The electromechanical actuators came from a previous NASA–Air Force–Navy research program. The parafoil was from an Army program to deliver heavy payloads within 500 feet of a target. Because the parafoil can flare, the vehicle doesn't have to be stressed to withstand a high-speed landing.

The installation work on the first X-38, called Vehicle 131, was completed in late May 1997, and the vehicle was returned to Dryden on June 4. The first captive flight aboard the B-52—the same aircraft that conducted the drop tests of the original lifting bodies—was made on July 31, 1997. Because of technical problems and winter weather, the first free flight was delayed into the spring of 1998, to March 12. At 7:29 A.M. PST, the B-52 rolled down the Edwards runway and lifted off with a deafening roar, its eight engines billowing black smoke. For the first time in nearly a quarter of a century, a new lifting body was about to make its first flight. The B-52 made a practice run over the drop zone, then began the final pass. The X-38 was to land on a section of brush-covered desert, with a crossroads at the planned touchdown point. With the low touchdown speed of the parafoil, the X-38 could land off-runway, in any flat, open field. At 8:30, the countdown reached zero, and the X-38 fell free of the B-52 pylon. The deployment sequence began, and the X-38 was soon descending slowly below the red, white, and blue parafoil. During the 9-minute flight, the X-38 made several turns under ground control. It crossed over a dirt road, then touched down in the desert. The vehicle skidded about its own length before coming to a stop. The initial test of the lifting body and parafoil combination had proved successful.

The free-flight tests should continue through late 1999. The next step would be a space test of an unmanned X-38. The vehicle would be carried into orbit by the space shuttle, then would be released to make a reentry and landing. This would be much like the 1967 proposal for an orbital M-2 flight aboard a Saturn 1B. The CRV would go into operation aboard the space station in the first years of the twenty-first century.

Initial estimates projected that the cost of the development program through completion of two space test vehicles could be less than $80 million. The original development cost to build a capsule-type CRV in the mid-nineties was more than $2 billion. Final costs of the program, including four operational vehicles, would be less than a quarter of that. A lifting-body design had added flexibility over a capsule CRV. Because of a lifting body's cross-range maneuvering capability, a desert landing site in the United States, Kazakhstan, or Australia would be in reach every 4.5 hours, versus as long as 18 to 20 hours for a capsule. The requirements for that added time in orbit would make the capsule CRV a much more complex spacecraft.

Like the HL-20 concept, the CRV could also serve as a crew transport vehicle. A proposal has been made to use the CRV as the basis of a joint U.S.-European spacecraft for launch on the French Ariane 5 booster. Its present configuration, however, is solely as a lifeboat, with the onboard life support system limited to 9 hours of operation after separation from the space station.

SINGLE-STAGE-TO-ORBIT LAUNCH VEHICLES

Since the early years of the space age, one goal has been to build a single-stage-to-orbit (SSTO) launch vehicle. Such a vehicle could take off from a launch pad or runway, go into orbit, then return to Earth without discarding any stages or fuel tanks. The idea would be to conduct airplanelike operations in space. The vehicle would only have to be serviced and fueled before it could again return to space.

In the early sixties, the Air Force undertook studies of such a vehicle, called the aerospace plane. The effort ran into severe technical problems and was canceled while still in the study phase. A conventional booster uses several stages to keep its weight down. As fuel aboard the first stage is burned, the now-empty tanks become excess weight. By separating the first stage, the useless weight of the tanks and engines is eliminated. A smaller, lighter second stage ignites and carries the payload toward orbit. For an SSTO vehicle to be successful, the mass fraction ratio (the total propellant weight divided by the total vehicle weight) must be around 0.91. With early sixties-vintage materials, that was not possible. By the mid-eighties, new carbon composite materials had been developed that seemed to make an SSTO vehicle feasible. The loss of space shuttle *Challenger* also sparked interest in a successor. In the late eighties, one emerged as the X-30 National Aero-Space Plane (NASP) program.

The X-30 was a large manned vehicle with a lifting-body fuselage. The comment was made that it should be called the M2-F4, as it had the flat top, rounded undersides, and twin fins of the M-2 shape. The vehicle was powered by a combination ramjet-scramjet (supersonic combustion ramjet) mounted under the fuselage. As the X-30 climbed, the engine would burn atmospheric oxygen with its liquid hydrogen fuel, improving the mass fraction ratio, as the vehicle did not have to carry oxygen.

The X-30 program soon ran into technical problems. It tried to integrate many new and untested technologies into a single full-scale vehicle. It was not possible to ground-test scramjet designs; actual flight tests above Mach 8 were needed to validate their designs (much as wind tunnels could not provide data on transonic and supersonic speeds in the late forties, requiring development of the X-1).

In an interview shortly before his death, Milt Thompson noted that the X-30 program had been sold on the basis that it could be developed for $3 billion and that the technology already existed. He said it was clear that all the technology did not exist and that the price had nearly tripled. He felt the financial problems of the United States were such that we could not afford such a vehicle, and he hoped that some type of manned hypersonic vehicle, austere and smaller than the X-30, could be built.

In June 1993, the X-30 vehicle was canceled. Its replacement was the Hypersonic Flight Test Experiment (HYFLTE) program, which would use Minuteman 2 ICBMs to boost unmanned subscale models to speeds of Mach 12 to 15. In November 1994, the NASP program was ended. The basic problem of the X-30 had been that it tried to put several new technologies, such as large scramjet motors, that were not yet ready for full-scale flight testing into a large vehicle. Over the next several years, there was a reassessment of the problems of both hypersonic flight and SSTO vehicles.

The different parts of the X-30 program were split off into several different efforts that would test the different technologies in a step-by-step manner, rather than in one huge leap. The hypersonic element of the program was reoriented into a simpler version of the HYFLTE program. This emerged as the Hyper-X program, which was to demonstrate technologies for air-breathing hypersonic flight.

Four flights were planned, one each to Mach 5 and 7, with two more at Mach 10. The Hyper-X vehicle is 12 feet long with a 5-foot wingspan and resembles a much smaller version of the X-30. Carried on the nose of a Pegasus rocket, it would be launched at about 40,000 feet from the NASA B-52. The Pegasus would boost the Hyper-X to the test speed at about 100,000 feet. It would then separate and begin powered flight. The ground track would extend over the Pacific Ocean for nearly 400 miles. The Hyper-X would come down near San Nicholas Island and not be recovered.

The SSTO element of the X-30 program became a technological demonstrator to test a small unmanned suborbital vehicle before building an operational SSTO version. On August 5, 1994, President Bill Clinton signed a space policy document calling for a major reduction in launch costs. Current costs run between $5,000 and $10,000 per pound. The policy statement called for a competition to produce a new SSTO test vehicle, the X-33, which could then be developed into a replacement for the shuttle in 2006 to 2012. On October 19, 1994, NASA issued a request for proposals. Less than six months later, on March 8, 1995, NASA reduced the X-33 contenders to three—Lockheed Martin, McDonnell Douglas, and Rockwell International.

The three vehicles were very different. The Lockheed Martin design was a wedge-shaped lifting body with two tip fins, resembling the HL-10. It used linear aerospike engines, rather than conventional rockets with bell nozzles. The McDonnell Douglas DC-Y design was a cone-shaped, vertically launched and landing vehicle. It was based on the DC-X test vehicle flown in 1993 and 1994. After reentry, it would restart its engines and make a soft landing. The Rockwell International proposal was based on its shuttle experience. It used a cylindrical fuselage with wings and two fins at the rear.

Over the next year, the three designs and their business plans were refined. Unlike the shuttle, they were to be commercial launch vehicles. Although

NASA would provide design requirements and funding, the contractors would have to invest large amounts of their own money, then operate the final SSTO vehicles.

On July 2, 1996, Vice President Al Gore announced that the Lockheed Martin proposal for the X-33 had been selected. There were several factors affecting the decision. The DC-Y proposal was too advanced and risky, while the Rockwell design was too shuttlelike. The Lockheed Martin design's aerospike engines combine both advanced and proven technology. The heat shield is a metallic material developed as part of the classified Have Region program. The Lockheed Martin design also offers the advantage of being built at the Skunk Works, home of the U-2, A-12, and F-117. It is a small group with a history of building technically innovative aircraft on tight schedules and under budget. The design has the lightest weight, with the lowest reentry heating of the three proposals, part of the heritage of the lifting bodies. As Dale Reed noted, "They're using it because of the confidence we built into it years ago. Lockheed engineers said they would not have come up with that shape if we hadn't done that work."

The X-33 is to be a half-scale version of the final SSTO vehicle. Launches are scheduled to begin in 1999 from a facility at Haystack Butte, on the eastern portion of Edwards Air Force Base. The initial midspeed flights would land at Michael Army Air Field on the Dugway Proving Grounds in Utah. The final high-speed and long-range flights would conclude at Malmstrom Air Force Base, near Great Falls, Montana. The X-33 vehicle would make a horizontal landing, like the earlier lifting bodies. After completion of the flights, it would be returned to Edwards on the back of the 747, in a manner similar to the space shuttle. The maximum speed would be Mach 15, with a peak altitude of 250,000 feet. A total of fifteen flights would be made.

The X-33 test program would qualify the technology necessary to build a full-scale SSTO vehicle. Another aspect will be the testing of the ground servicing plan. A simplified operations plan is to be developed, using minimum ground support staff rather than the standing army previously required. The hardware is to be designed to meet the servicing plan, rather than developing a servicing plan to fit the hardware.

Once the test program was completed, a decision could be made on the next step, the Venture Star. The SSTO vehicle would be 127 feet long, with a span over the tip fins of 128 feet. Payload would be 59,000 pounds. The all-important payload cost would be, according to Lockheed Martin estimates, on the order of $600 per pound. The Venture Star could be ready for another flight a week after returning from space, much faster than the three-month turnaround for the shuttle. Once fully operational, the Venture Star could provide a savings of $3 billion per year in launch costs.

The Venture Star's payload bay has fittings to accommodate all existing or planned satellites and upper stages. The Venture Star could also accommodate manned satellite servicing flights and crew transfers to the space station. The crew would ride in a pressurized module within the payload bay. If existing plans are carried out, the Venture Star would begin operations in 2004, ushering in the second century of heavier-than-air flight.

A SHORT VOYAGE TO A DISTANT STAR

That first morning, a long lifetime ago, dawned bitterly cold. A frigid north wind froze puddles of water in the sand dunes of Kitty Hawk. For most of the morning, Orville and Wilbur Wright waited for the wind to die down. Both they and their machine were ready, and at about 10 A.M., they decided to try. A flag summoned members of the local lifesavers station. By 10:30 A.M., the launch rail was complete and the airplane's engine was running. Orville climbed aboard the lower wing, while Wilbur steadied the right wingtip.

At 10:35 A.M., December 17, 1903, Orville shifted the release lever, and the airplane began moving down the rail. Wilbur ran alongside. After a 40-foot takeoff roll, the first airplane lifted off to test the winds. One of the lifesavers, John T. Daniels, snapped the shutter of the Wrights' camera, freezing that instant forever. The photograph caught Wilbur in midstride, looking, as if in amazement, at what the brothers had created. The airplane, the *Wright Flyer,* is suspended above the ground, its wings a bright white in the weak December sunlight. In the distance, sand and sky meet, becoming a dim and misty horizon, beckoning them onward, ever onward.

Four flights were made that day toward that horizon. The last covered 852 feet and took 59 seconds. Without question, sustained heavier-than-air powered flight had been achieved.

Orville Wright lived another forty-four years. In that time, he saw wood and fabric biplanes give way to all-metal monoplanes, propellers replaced by jet engines, and airliners spanning the world. Three months before Wright's death, Chuck Yeager made the first supersonic flight in the X-1. When Wright died on January 30, 1948, he was eulogized as "a man who was just one of folks like us—middle class, mid-Western American." But as he was laid to rest, aviation was on the verge of a revolution. The next two decades would see tremendous advances. The frontier was not just the sky, but also space. How far that frontier extended was not clear in 1948, most of all to a seventeen-year-old boy named Neil A. Armstrong.

The dream of human flight is as old as history, but it was only in our time that the dream was made real. In a brief span of years, we have gone from Kitty Hawk to the Moon, and from the *Wright Flyer* to the space shuttle. How have we come so far so fast? It is because of men of vision who have the courage to dream and the determination to make those dreams a reality in the face of hardships and setbacks. The *Wright Flyer* and the lifting bodies and the space shuttle are all expressions of that same dream and determination. They come from different times and embody the different technologies of wood and fabric, aluminum and plastic, electronics and heat-resistant tiles, but each is an answer to the call of the beckoning horizon.

As we near the start of a second century of flight, we also face hardships and setbacks. The future holds new technologies and new challenges yet unknown. As always, the horizon beckons us onward, ever onward, toward some distant star.

BIBLIOGRAPHY

DiGregorio, Barry E. "Test Pilot Elite." *Quest,* Spring 1994, 13–20.

Dorr, Les Jr. "NASA's Hot Rods: Souped-up Coups Served Role in Space Race." *Quest,* Spring 1993, 12–15.

Gauthier, Daniel James. "Shuttling into the New Century." *Quest* 5 no. 2 (1996): 32–33.

Hallion, Richard P. *Supersonic Flight: Breaking the Sound Barrier and Beyond.* London: Brassy's, 1997.

———. *Test Pilots: The Frontiersmen of Flight.* Washington, D.C.: Smithsonian Institution Press, 1988.

———. *On the Frontier.* Washington, D.C.: NASA Scientific and Technical Information Branch, 1984.

Hengeveld, Ed. "Land Landings for Gemini." *Quest,* Fall 1993, 4–13.

Houchin, Roy F. II. "Why the Air Force Proposed the Dyna-Soar X-20 Program." *Quest,* Winter 1994, 5–12.

———. "Why the Dyna-Soar X-20 Program Was Canceled." *Quest,* Winter 1994, 35–37.

Kempel, Robert W., Weneth D. Painter, and Milton O. Thompson. *Developing and Flight Testing the HL-10 Lifting Body: A Precursor to the Space Shuttle.* Edwards, Calif.: Dryden Flight Research Center, 1993.

Miller, Marcianne, and John Terreo. *Legacy of the Lifting Bodies.* Edwards Flight Test Center: Computer Sciences Corp., 1995.

Peebles, Curtis. *High Frontier: The United States Air Force and the Military Space Program.* Washington, D.C.: Air Force History and Museum Program, 1997.

———. "The Origins of the U.S. Space Shuttle, 1." *Spaceflight,* November 1979, 435–442.

———. "The Origins of the U.S. Space Shuttle, 2." *Spaceflight,* December 1979, 487–492.

Pesavento, Peter. "Russian Space Shuttle Projects, 1957–1994: Part 1." *Spaceflight,* May 1995, 158–161.

———. "Russian Space Shuttle Projects, 1957–1994: Part 2." *Spaceflight,* June 1995, 192–195.

———. "Russian Space Shuttle Projects, 1957–1994: Part 3." *Spaceflight,* July 1995, 226–229.

———. "Russian Space Shuttle Projects, 1957–1994: Part 4." *Spaceflight,* August 1995, 264–266.

Reed, R. Dale, with Darlene Lister. *Wingless Flight: The Lifting Body Story.* Washington, D.C.: NASA History Office, 1997.

Rotundo, Louis. *Into the Unknown: The X-1 Story.* Washington, D.C.: Smithsonian Institution Press, 1994.

Smith, Terry. "The Dyna-Soar X-20: A Historical Overview." *Quest,* Winter 1994, 13–18.

Thompson, Milton O. *At the Edge of Space: The X-15 Flight Program.* Washington, D.C.: Smithsonian Institution Press, 1992.

Vitelli, John. *The Start Program and the SV-5 Configuration.* SAMSO Historical Division, 1967.

Wilkinson, Stephan. "The Legacy of the Lifting Body." *Air and Space,* April–May 1991, 50–62.

Yeager, Chuck, and Leo Janos. *Yeager: An Autobiography.* New York: Bantam Books, 1985.

INDEX

NOTE: This index was made by Curtis Peebles. Entries for Aircraft; Lifting Bodies; and Spacecraft and Satellite Programs precede the start of the alphabetic sequence.

Alphabetical index follows listing of aircraft, lifting bodies, and spacecraft.

Alphabetical index follows listing of aircraft, lifting bodies, and spacecraft.

Alphabetical index follows listing of aircraft, lifting bodies, and spacecraft.

Alphabetical index follows listing of aircraft, lifting bodies, and spacecraft.

E

F

G

Alphabetical index follows listing of aircraft, lifting bodies, and spacecraft.

Alphabetical index follows listing of aircraft, lifting bodies, and spacecraft.

Alphabetical index follows listing of aircraft, lifting bodies, and spacecraft.

Alphabetical index follows listing of aircraft, lifting bodies, and spacecraft.

R

S

Alphabetical index follows listing of aircraft, lifting bodies, and spacecraft.

Alphabetical index follows listing of aircraft, lifting bodies, and spacecraft.

W

X

Y

Z

Alphabetical index follows listing of aircraft, lifting bodies, and spacecraft.